Computer Architecture and Design Methodologies

Twilight zone of Moore's law is affecting computer architecture design like never before. The strongest impact on computer architecture is perhaps the move from unicore to multicore architectures, represented by commodity architectures like general purpose graphics processing units (gpgpus). Besides that, deep impact of application-specific constraints from emerging embedded applications is presenting designers with new, energy-efficient architectures like heterogeneous multi-core, accelerator-rich System-on-Chip (SoC). These effects together with the security, reliability, thermal and manufacturability challenges of nanoscale technologies are forcing computing platforms to move towards innovative solutions. Finally, the emergence of technologies beyond conventional charge-based computing has led to a series of radical new architectures and design methodologies.

The aim of this book series is to capture these diverse, emerging architectural innovations as well as the corresponding design methodologies. The scope covers the following.

- Heterogeneous multi-core SoC and their design methodology
- Domain-specific architectures and their design methodology
- Novel technology constraints, such as security, fault-tolerance and their impact on architecture design
- Novel technologies, such as resistive memory, and their impact on architecture design
- Extremely parallel architectures

Siyi Wang • Anupam Chattopadhyay

Quantum Arithmetic

Unlocking the Future of Computing

Springer

Siyi Wang
College of Computing and Data Science
Nanyang Technological University
Singapore, Singapore

Anupam Chattopadhyay
College of Computing and Data Science
Nanyang Technological University
Singapore, Singapore

ISSN 2367-3478 ISSN 2367-3486 (electronic)
Computer Architecture and Design Methodologies
ISBN 978-981-95-6038-7 ISBN 978-981-95-6039-4 (eBook)
https://doi.org/10.1007/978-981-95-6039-4

This Springer imprint is published by the registered company Springer Nature Singapore Pte Ltd.
The registered company address is: 152 Beach Road, #21-01/04 Gateway East, Singapore 189721, Singapore

Acknowledgements

This book is the result of my work as a PhD student at Nanyang Technological University, Singapore. I would like to take this opportunity to express my heartfelt gratitude to everyone who has supported me throughout my PhD journey. Working in the field of quantum computing, a complex and advanced area of computer science, has been challenging. The early stages of my research were particularly difficult, but thanks to the support and guidance of those around me, I was able to successfully complete this book.

First, I am deeply grateful to my supervisor, Prof. Anupam Chattopadhyay, for his unwavering support, valuable insights, and patience over the years. I will never forget our many in-depth discussions at Canteen 2 and North Spine, which provided clarity and direction to my research. His mentorship has been a cornerstone of my work, and I will always be profoundly thankful.

I also want to sincerely thank my colleagues and collaborators, whose contributions have greatly enhanced the quality of my research. The acceptance of our papers in top-tier journals and conferences highlights the success of our collaboration. I am especially proud that our work received the only Best Paper Award at the 2024 IFIP/IEEE International Conference on Very Large Scale Integration. I also would like to thank my family and my friends for their support throughout this journey. Their encouragement and belief in me have been a constant source of motivation.

Lastly, I want to express my gratitude to myself. Throughout this journey, I have never given up and have consistently supported and encouraged myself, especially during difficult moments that were hard to share with others. The past version of Siyi was always by my side, offering strength, care, patience, and love when I needed it most. These experiences have helped me become more independent and resilient, better care for and love myself, and ultimately establish a stable, lifelong friendship with myself.

With all my heart, thank you.

Singapore, Singapore
July 2025

Siyi Wang

About the Book

In recent decades, quantum computing has witnessed remarkable progress, achieved through various advancements across the technology stack, such as quantum error-correcting codes and quantum algorithms. It is noteworthy that several promising algorithms incorporate quantum arithmetic circuits as fundamental building blocks. However, the inherent properties of quantum circuits challenge the efficiency of quantum arithmetic circuits, prompting creative solutions. While several quantum arithmetic circuits have been proposed, there is still a substantial gap in quantum arithmetic circuits optimization. For example, methods for integrating classical techniques such as high radix strategy into quantum designs still remain underexplored, highlighting a significant space for further advancement.

This book aims to improve the performance of quantum arithmetic designs by integrating classical arithmetic circuit design techniques. Specifically, this book contains the following research: (1) The utilization of classical higher radix strategy to enhance quantum addition, reducing overall cost by lowering T-depth and T-count. The proposed design outperforms existing quantum carry-lookahead adders. (2) The application of Ling structure, achieving a reduced Toffoli-depth of $4\log(n/2)$. (3) The exploration of the Kogge-Stone architecture, resulting in the lowest Toffoli-depth of $\log(n)$ for quantum adders. (4) The development of an efficient modular addition framework using carry-save architecture, demonstrating a 33.3% improvement in Toffoli-depth compared to existing designs. (5) The design of an optimized quantum non-restoring divider framework. (6) A comprehensive overview of current advancements in quantum arithmetic designs, offering insights into potential research directions. These proposed designs are mapped to real-world quantum hardware architectures, including Noisy Intermediate-Scale Quantum systems, to validate their functionality and evaluate efficiency. Their applications in quantum machine learning and cryptography further underscore the broader impact of these improvements.

Overall, this book makes significant advancements in reducing quantum circuit depth and improving resource efficiency, contributing to the development of scalable and efficient large-scale quantum circuits. By adapting classical design strategies to quantum circuits, it not only enhances various quantum circuit designs but also fosters further exploration at the intersection of classical and quantum computing.

Contents

About the Authors

Dr. Wang Siyi is a research fellow at Nanyang Technological University (NTU), Singapore, under the supervision of Prof. Anupam Chattopadhyay. Her research focuses on quantum computing and cybersecurity, with notable contributions including MalaQ, the first malware targeting ion-trapped quantum computers. She has published extensively on quantum arithmetic, cryptanalysis, and quantum cybersecurity, and her work has been recognized with several awards, including Best Student Presentation at Yale's Quantum Computer Cybersecurity Symposium, Best Paper at the IFIP/IEEE VLSI-SoC Conference, and Best Poster at AsiaCCS 2025.

Anupam Chattopadhyay received his BE degree from Jadavpur University, India, MSc from ALaRI, Switzerland, and PhD from RWTH Aachen in 2000, 2002, and 2008, respectively. Presently, he is appointed as an associate professor (with tenure) in the College of Computing and Data Science, NTU. His research in the domains of Electronic Design Automation, Post-quantum Cryptography, AI Security, and Secure IoT has led to multiple spinoffs and technology transfers. He has served in the Technical Program Committee (TPC) of major Electronic Design Automation (EDA)/Very Large Scale Integrated Circuits (VLSI) conferences and edited a Springer Major Reference Work on Computer Architectures. His research has garnered several best paper awards, most recently in VLSI-SoC 2024 for Quantum adders, and in CHES 2025 for side-channel attack on PQC standard Kyber.

Acronyms

ABC	Atanasoff-Berry Computer
BER	Binary Exponent-based Recombination
CBC	Cipher Block Chaining
CFB	Cipher Feedback
CLA	Carry Look-ahead Adder
CSA	Carry Save Adder
DDS	Direct Digital Synthesizer
DoS	Denial of Service Attack
DSE	Design Space Exploration
ECB	Electronic Codebook
ELU	Exponential Linear Unit
EPLG	Error per Layered Gate
FFT	Fast Fourier Transform
FPGA	Field Programmable Gate Array
GELU	Gaussian Error Linear Unit
IC	Integrated Circuit
ICS	Industrial Control Systems
IoT	Internet of Things
IV	Initialization Vector
LSI	Large-Scale Integration
MCC	Manchester Carry Chain
ML	Machine Learning
NISQ	Noisy Intermediate-Scale Quantum System
PC	Personal Computer
PQC	Parameterized Quantum Circuits
QAOA	Quantum Approximate Optimization Algorithm
QC	Qubit-Count
QECC	Quantum Error-Correcting Codes
QFT	Quantum Fourier Transformation
QMbead	Quantum Multiplier based on Exponent Adder
QML	Quantum Machine Learning

QNN	Quantum Neural Network
QPCA	Quantum Principal Component Analysis
QPE	Quantum Phase Estimation
QPU	Quantum Processing Unit
QSVM	Quantum Support Vector Machines
RCA	Ripple Carry Adder
ReLU	Rectified Linear Unit
STO	Search with Two Oracles
TC	Toffoli-Count
TD	Toffoli-Depth
TRADIC	Transistor Digital Computer
VLSI	Very-Large-Scale Integration
VQE	Variational Quantum Eigensolver

List of Figures

List of Tables

Chapter 1
Introduction

> *Quantum mechanics describes nature as absurd from the point of view of common sense. And yet it fully agrees with experiment. So I hope you accept nature as She is.*
>
> –Richard P. Feynman

1.1 Book Background

Quantum computing marks one of the most significant paradigm shifts since the advent of classical computers. Built upon the principles of quantum mechanics, it introduces novel computational methods that have the potential to surpass classical computing in both efficiency and capability. Unlike classical computers, which use binary bits, quantum computers rely on qubits, which must operate under strictly controlled physical conditions. By exploiting quantum phenomena such as superposition and entanglement, quantum computers can utilize qubits to process information in fundamentally different ways. These unique properties enable quantum computers to solve various problems that are currently infeasible for even the most advanced classical systems. Therefore, quantum computing has the potential to revolutionize various industries such as cryptography, drug discovery, and materials science.

Quantum circuits serve as the computational backbone of quantum algorithms, comprising qubits and quantum gates. Unlike classical bits, which exist in a definite state of either 0 or 1, qubits can exist in a superposition of both states simultaneously. Quantum gates—such as the Hadamard, CNOT, and Toffoli gates—are used to manipulate these quantum states and perform basic operations in quantum computing.

Classical computing, based on Boolean logic [22] and binary bits, has served as the foundation for significant technological progress in fields such as engineering, physics, and finance. In this paradigm, information is stored in binary bits, each of which is represented as either a 0 or a 1. Innovations such as the von Neumann architecture [219] and Moore's law [142], which predicted the exponential growth of transistor density, have been instrumental in the rapid advancement of

S. Wang, A. Chattopadhyay, *Quantum Arithmetic*, Computer Architecture and Design Methodologies, https://doi.org/10.1007/978-981-95-6039-4_1

computational power in classical computing. However, as researchers approach the physical limits of miniaturization and parallelism, classical computing is now facing fundamental barriers in efficiency. Quantum computing offers a path beyond the limitations of classical computing. The idea of a quantum computing was first proposed by Richard Feynman [53] in 1982. In 1985, David Deutsch[1] further formalized the concept of a universal quantum computer and proposed the first universal quantum Turing machine. In the 1990s, the development of algorithms such as Shor's factoring algorithm [193] and Grover's search algorithm [74] demonstrated that quantum computing could achieve exponential and quadratic speedups for specific tasks compared to classical computing.

In the novel and rapidly evolving field of quantum computing, the optimization of quantum arithmetic circuits has garnered extensive research attention since the efficiency of these circuits largely determines the computational capability and overall cost of numerous quantum algorithms and large-scale quantum systems. Important topics in the quantum arithmetic domain include the design of quantum adders, multipliers, and dividers, which serve as foundational components of various quantum circuits. It is essential to explore more efficient and robust quantum arithmetic designs, as they play a crucial role in the realization of computationally demanding tasks such as modular exponentiation, thereby improving the efficiency of relevant large-scale quantum algorithms, such as Shor's algorithm [193].

1.1.1 Basic Quantum Logical Gates

Here, we provide a concise overview of these basic quantum logic gates,[2] including corresponding quantum circuits as shown in Fig. 1.1.

1.1.2 Research Problem

With the ongoing progress of quantum computers, the research of quantum computing has shifted from theoretical contemplation to practical implementation and application. In this context, the efficiency of quantum arithmetic circuits has gained increasing concern. In general, T-depth, T-count, and qubits count are mainly used to measure the efficiency of quantum arithmetic circuits. T-depth presents the time cost for the overall circuit, T-count counts the total number of T operations, and qubits count demonstrates the number of qubits needed for the whole structure.

---

[1] https://plato.stanford.edu/entries/qt-quantcomp/

[2] While we do not delve into their definitions in this book, readers unfamiliar with these gates are encouraged to check Nielsen and Chuang's standard quantum computing textbook https://almuhammadi.com/sultan/books_2020/Nielsen_Chuang.pdf.

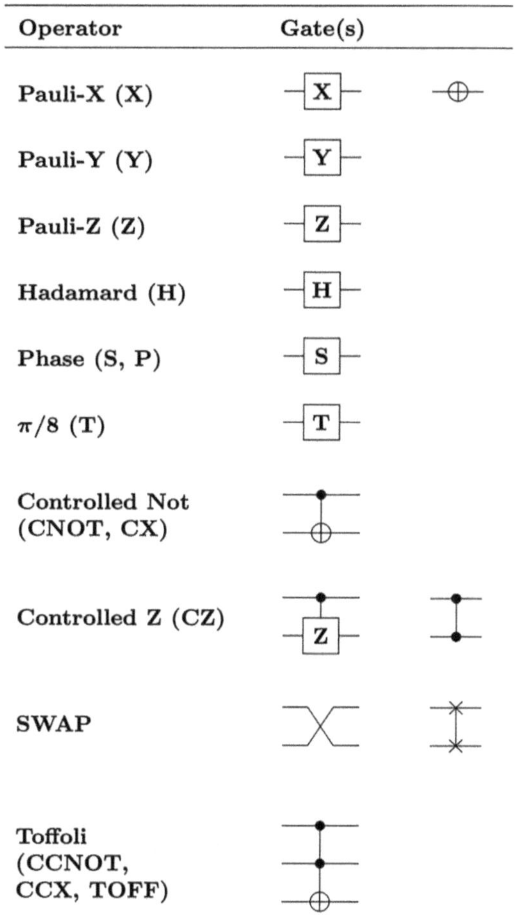

Fig. 1.1 Basic quantum logic gates

Quantum arithmetic circuits serve as the cornerstone of both quantum circuits and algorithms, demanding innovative strategies for efficiency improvement. Currently, there are three main directions to enhance efficiency: the reduction of T-count, T-depth, or qubits count. The primary emphasis of our research focuses on the reduction of T-depth, which is very important since it highly relates to the overall time complexity.

1.1.3 Gap and Motivation

Considering the interaction between TC, TD, and QC, it is important to find an efficient strategy for cost reduction and trade-off management in this research domain.

One fundamental challenge lies in the no-cloning theorem—the non-copyability of qubits—which prevents direct reuse of intermediate results and complicates the translation of classical arithmetic structures into quantum circuits. As a result, classical logic such as carry propagation and fan-out operations must be carefully restructured for quantum settings. To address this, we modified the classical structures, maximizing parallel structures while considering quantum circuit properties.

Beyond non-copyability, several other critical challenges further complicate quantum arithmetic design as follows:

- **Reversibility.** Quantum circuits must be fully reversible, unlike many classical arithmetic designs. This adds complexity to the logic and often increases the number of ancilla qubits and gates required.
- **Error Sensitivity.** Quantum gates are highly error-prone, and quantum states are fragile due to decoherence.
- **Limited Qubit Connectivity.** Real-world quantum devices typically have limited connectivity, requiring additional SWAP operations that can increase depth and error rates.
- **Measurement and Reuse Overhead.** Unlike classical logic, quantum circuits are constrained in their ability to measure and reuse qubits, especially on current quantum hardware.

In response to these challenges, this book proposes quantum arithmetic designs that aim to reduce Toffoli-Depth, while maintaining low Toffoli-Count and qubit usage. By tailoring building blocks to quantum constraints—such as using CNOT to realize repeat logic—we develop efficient, scalable arithmetic modules that advance both theoretical understanding and practical realizability.

Improving the efficiency of quantum arithmetic circuits is essential not only for enhancing specific applications like Shor's algorithm, but also for accelerating the advancement of the quantum computing revolution.

1.2 Research Aim and Objectives

The primary aim of this book is to enhance the performance of quantum arithmetic circuits by integrating classical design techniques. To achieve this aim, we organize our objectives around three coherent themes: (1) efficient arithmetic design, (2) practical hardware mapping and validation, and (3) application-driven exploration. Specifically, the objectives are as follows:

- **Design Optimization of Quantum Addition.** Investigate innovative methodologies to optimize resource requirements and time costs in the quantum addition domain, focusing on higher radix strategies [224], Ling structures [225, 226], and Slansky architectures [232].
- **Design Optimization of Quantum Modular Arithmetic.** Propose an optimized modular addition framework based on the carry-save architecture. This frame-

work targets significant improvements in time requirement compared to existing baselines.

- **Design Optimization of Quantum Division.** Extend optimization techniques in quantum addition by proposing an efficient quantum non-restoring divider framework to further enhance performance in quantum division operations.
- **Survey and Synthesis.** Provide a systematic overview of quantum arithmetic circuit designs, highlighting common techniques, performance trade-offs, and open challenges to guide future research directions.
- **Hardware Mapping and Validation.** Map the proposed designs to real-world quantum hardware architectures, including noisy intermediate-scale quantum (NISQ) systems, to validate functionality and assess their efficiency.
- **Security Implications.** Analyze the potential vulnerabilities of quantum system by designing and simulating a quantum malware MalaQ tailored for ion-trapped quantum computers. This work highlights the significance of quantum cybersecurity.
- **Applications in Emerging Domains.** Explore the role of efficient quantum arithmetic in high-impact application domains such as quantum machine learning and quantum cryptanalysis.

1.3 Evaluation Metrics

In Fig. 1.2, the architecture of a generic quantum system is presented in detail. The evaluation of the quantum system is conducted comprehensively using various metrics, which are described below for systematic understanding:

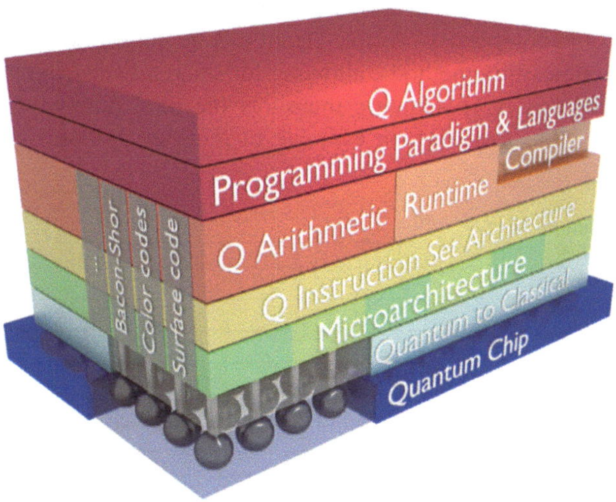

Fig. 1.2 Quantum computers' architecture [56]

- **Quantum Volume [39].** It represents the computational space complexity of quantum systems. However, for small systems, it may not offer a holistic view, and for larger systems, quantum volume experiments may become impractical for classical simulation.
- **CLOPS [220].** This is a speed metric evaluating the efficiency of processors in executing quantum volume circuits in series. In brief, CLOPS integrates both quantum and classical computing, providing a holistic measure of system speed.
- **Layer Fidelity [140].** It provides a detailed understanding of quantum systems, capturing their capability to execute current user circuits. Hence, layer fidelity offers insights into individual qubits, gates, and crosstalk, serving as a comprehensive benchmark for varying circuit sizes.
- **Error per Layered Gate (EPLG) [17].** This is a quantum metric calculated as $EPLG = 1 - LayerFidelity^{\frac{1}{N-1}}$, providing a refined measure of error by considering layer fidelity and its impact on overall performance.

In this book, we focus on the most relevant metrics. Specifically, we emphasize two main categories of designs. For each category, we utilize corresponding metrics to evaluate the efficiency of the arithmetic designs, as outlined below.

1.3.1 Evaluation Metrics for Clifford+T-Based Design

For Clifford+T-based design, we focus on three key metrics. Figure 1.3 illustrates a basic example of calculating the primary quantum circuit metrics used in this book. These metrics are outlined as follows:

- **Toffoli-Depth (TD):** It is the most important quantum cost metric as it represents the number of computational layers that include Toffoli gates, which relates to the overall time complexity. Toffoli-Depth can be converted into T-depth using specific decomposition methods, allowing for standardized evaluation.
- **Toffoli-Count (TC):** This metric measures the total number of Toffoli gates within the circuit, providing an estimate of the gate complexity and quantum

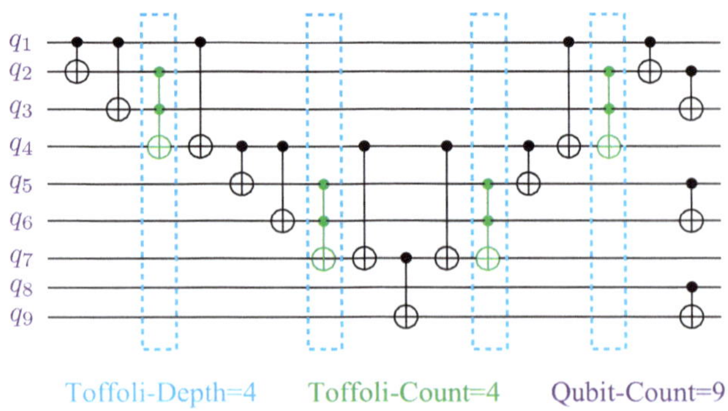

Fig. 1.3 Basic quantum evaluation metrics for Clifford+T-based designs. The dotted boxes highlight the Toffoli gates in the circuit. For the circuit shown, the qubit count is 9, Toffoli-Count is 4, and Toffoli-Depth is 4

resource consumption. It can be converted into T-count using specific decomposition methods for standardized comparisons.

- **Qubit Count (QC):** This metric denotes the total number of qubits required in the specific quantum circuits, correlating with the size of the quantum circuit.

1.3.2 Evaluation Metrics for QFT-Based Design

For QFT-based design, the elementary gate set consists of the Hadamard gate and the controlled phase gate ($CP(\phi)$). Controlled phase gates form a family of controlled rotational operations over the z-axis on the Bloch sphere, such that $CP(\phi) = |0\rangle\langle 0| \otimes I + |1\rangle\langle 1| \otimes [|0\rangle\langle 0| + e^{i\phi}|1\rangle\langle 1|]$. The identity gate occurs when $\phi = 0$, while the controlled Z gate occurs when $\phi = \pi$. Both are in the Clifford set. For non-Clifford gates, such as the controlled S gate when $\phi = \pi/2$, the controlled T gate when $\phi = \pi/4$, and the generally controlled phase gate when the rotation angle is arbitrarily set, more effort is required for experimental realizations. Thus the performance of QFT-based arithmetic circuits depends predominantly on the number of non-Clifford controlled phase gates. In Fig. 1.4, we illustrate a 3-qubit QFT-based quantum adder, which has two QFT blocks and three non-Clifford controlled phase gates for addition. The total number of non-Clifford controlled phase gates is 9. To characterize the overall performance and complexity of QFT-based arithmetic algorithms, we summarize the following three metrics:

- **QFT-Count:** Due to the fixed structure of QFT and its inverse, which depend only on the qubit count, the depth and count of non-Clifford controlled phase gates are always quadratic to the qubit count. Therefore, we consider the number of QFT and its inverse as a performance metric. Typically, one block of QFT for an n-qubit circuit requires $n(n-1)/2$ non-Clifford controlled phase gates.
- **Non-Clifford-CP-Count:**After QFT, arithmetic operations such as summations and multiplications are performed on the Fourier basis and usually require

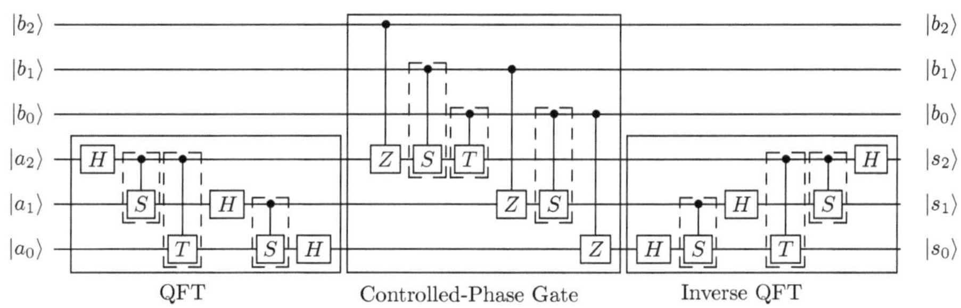

Fig. 1.4 Basic quantum evaluation metrics for QFT-based designs. For 3-qubit adder, the QFT-Count is 2 and the Non-Clifford-CP-Count is 3. Note that the Non-Clifford-CP-Count does not include controlled phase gates in the QFT blocks

sequences of additional controlled phased gates. Therefore, we consider the number of non-Clifford controlled phase gates, excluding those in the QFT subroutines, as another performance metric. A preferred algorithm typically has no additional controlled operations in the middle but is instead designed with single-qubit operations.
- **Qubit Count:** This metric is the same as in Clifford+T design.

Here, we point out that sometimes the evaluation metrics of QFT-based designs are connected to those of Clifford+T designs. For example, in Paler's work [158], the QFT-based metric of Non-Clifford-CP-Count is linked to the Clifford+T-based metric of Toffoli-Count by merging adjacent Hadamard gates and controlled rotation gates and formulating them with Clifford and Toffoli operations. Related work on the discussions about resource costs for QFT-based algorithms using Clifford+T-based metrics can be found in reference [115, 144, 162].

1.4 Organization

This book is structured to follow a logical progression that mirrors the main research aim and objectives outlined in Sect. 1.2. It begins by establishing foundational context and then proceeds to core design innovations, practical validation, and finally, real-world applications. Each chapter builds upon the previous to form a coherent and focused exploration of quantum arithmetic design and its implications:

- Chapter 2 describes the relevant background.
- After that, Chap. 3 presents our work on efficient quantum arithmetic designs, detailing six work that address key quantum arithmetic operations, including addition, modular addition, division, as well as a comprehensive overview of the current state of the art in quantum arithmetic.
- In Chap. 4, we explore the applications of quantum arithmetic designs, specifically within quantum machine learning and cryptanalysis domains.
- We conclude the book and discuss future research directions in Chap. 5.

The overall structure is intentionally modular to reflect the layered nature of the research: starting from circuit-level innovations, progressing through hardware-level considerations, and concluding with domain-level impact.

Chapter 2
Background

Study the past, if you would define the future.

–Confucius

This chapter reviews prior work in quantum arithmetic, starting with essential background information. It then discusses quantum architectures relevant for mapping our designs to quantum hardware and highlights the practical significance of quantum arithmetic through its applications.[1]

2.1 History of Computing

> For suppose that every tool we had could perform its task, either at our bidding or itself perceiving the need, and if-like the statues made by Dædalus or the tripods of Hephæstus, of which the poet says that 'self-moved they enter the assembly of the gods'-shuttles in a loom could fly to and fro and a plectrum play a lyre all self-moved, then master-craftsmen would have no need of servants nor masters of slaves.
>
> –Aristotle

The pursuit of automation and efficiency has been a fundamental driver of technological advancement throughout history. From ancient tools like the abacus to modern classical computers and emerging quantum technologies, each era of computing has pushed the boundaries of what is possible. In this section, we will explore the evolution of computing, starting with early human calculation methods, advancing through the development of classical computing, and culminating in the transformative paradigm of quantum computing.[2]

[1] This chapter includes our paper [229].

[2] For more details, one can refer to Ifrah's book: *The Universal History of Computing: From the Abacus to the Quantum Computer* [101].

S. Wang, A. Chattopadhyay, *Quantum Arithmetic*, Computer Architecture and Design Methodologies, https://doi.org/10.1007/978-981-95-6039-4_2

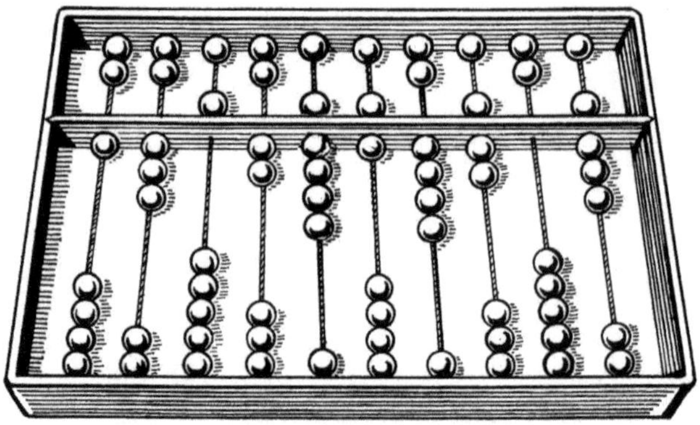

Fig. 2.1 Abacus. This is a bi-quinary coded decimal-like abacus that represents the number 1, 352, 964, 708

2.1.1 Human Calculation

In the earliest stages of computation, calculations relied entirely on human effort. Consequently, this process was not only slow and difficult but also monotonous due to its repetitive nature. To alleviate this burden and improve efficiency, early civilizations began developing tools to simplify arithmetic tasks. Among all of these innovative tools, the most representative was the abacus.

The term *abacus* was first recorded by Xu Yue (AD 160–220) in his work Suanshu Jiyi (Notes on Arithmetic Methods). As shown in Fig. 2.1, the traditional Chinese abacus featured a rectangular wooden frame containing vertical rods with movable beads. Each rod was split into two sections by a horizontal beam: The upper section typically held one bead representing the value of five, while the lower section contained four beads, each representing one. Calculations were executed by sliding these beads along the rods, allowing users to carry out arithmetic operations effectively.

2.1.2 Classical Computing

2.1.2.1 Mechanical Calculation

With the onset of the Industrial Revolution in the nineteenth century, commercial activity and international banking expanded rapidly, creating a growing demand for automated and efficient calculation (as shown in Fig. 2.2).

This surge in demand prompted the search for efficient and reliable solutions to perform calculations faster and more cost-effectively. One of the most significant early inventions was Pascal's adding machine, created in 1642, as shown in Fig. 2.3.

Fig. 2.2 Industrial Revolution

Fig. 2.3 Pascal type calculator [96]. In 1642, Blaise Pascal devised what was probably the first simple adding machine using geared wheels

Motivated by the tedium of calculating long tax lists for his father, Pascal designed a gear-based machine capable of adding numbers and carrying over values between places. However, the process of adding large numbers required carrying across multiple places, which demanded much greater force than could be provided by hand. Despite its limitations, Pascal's design marked a crucial step forward in automating arithmetic tasks. In the 1670s, Leibniz expanded upon this work by developing the "stepped reckoner," which could perform multiplication efficiently. Leibniz's machine used large, stepped drums meshing with gears to carry out calculations, representing a significant advancement in the field of mechanical calculation.

The growing demand for faster and more efficient calculation, driven by the Industrial Revolution, marked the beginning of the era of classical computing.

Overall, these early mechanical calculators laid the foundation for the development of digital computers and other advanced computing systems.

2.1.2.2 Classical Computer

The history of classical computing is marked by several significant milestones that led to the sophisticated classical computer we use today.

The *Vacuum Tube Computer Era* began with the invention of the audion in 1906 by Lee De Forest. The term "audion," an electronic detecting or amplifying vacuum tube, marked the beginning of the electronic revolution and laid the foundation for the development of the first generation of electronic computers, known as vacuum tube computers. This shift from mechanical to electronic systems was crucial for classical computing. One early example of this transition was the IBM 601 Multiplying Punch, which could perform multiplication. Although rudimentary by modern standards, it demonstrated the advantages of electronic over mechanical calculation. Building on these early attempts, in 1936, Alan Turing introduced the concept of the Turing Machine, an abstract model for computation that replaced human mathematical operations with a theoretical machine capable of performing calculations. This concept became a fundamental breakthrough in the theory of computing. One year later, John Vincent Atanasoff and his graduate student Clifford Berry developed the Atanasoff-Berry Computer (ABC), which was the first computer capable of solving linear algebraic equations, although it was non-programmable. Despite its limitations, the ABC significantly influenced the development of subsequent computers such as the ENIAC, Z1, Z2, and Z3. In the late 1940s, John von Neumann introduced the concept of the stored-program computer. This revolutionary idea allowed both the program and data to be stored in memory, forming the basis of modern computer architecture (as shown in Fig. 2.4).

Although vacuum tube computers marked the beginning of classical computing, they suffered from significant limitations, including large physical size, high power consumption, and frequent malfunctions. In the late 1940s, John Bardeen, Walter Brattain, and William Shockley at Bell Labs invented the transistor (Nobel Prize in Physics in 1956), offering a more efficient computing solution. This breakthrough laid the foundation for second-generation computers, known as *transistor computers*. In 1954, Bell Lab assembled the world's first transistor computer, named TRADIC (Transistor Digital Computer), which used 800 transistors. This marked a significant milestone in computing technology, demonstrating the potential of transistor-based computers. Just three years later, Seymour Cray and his team introduced the CDC 1604, the first fully transistorized supercomputer, offering greater reliability and efficiency. The development of transistors revolutionized computing, paving the way for smaller, faster, and more affordable computers (Fig. 2.5).

The third-generation computers were based on *integrated circuit (IC) technology*, which included microprocessors, memory, and input/output devices. While the transistor represented a significant improvement over vacuum tubes, it still gen-

Fig. 2.4 Electronic revolution

erated considerable heat, which could damage the sensitive components within the computer. In 1952, Geoffrey Dummer proposed integrating electronic components onto a single board to establish smaller and more energy-efficient computers, laying the groundwork for the IC technology. This concept was realized in 1958 by Jack Kilby, who created the first functioning IC. In 1960, Robert Noyce developed a more practical silicon-based IC that eliminated the need for external wiring. Therefore, Kilby and Noyce are both recognized as the fathers of the integrated circuit. By utilizing IC technology, IBM successfully developed the first general-purpose computer in 1964. Compared to the second-generation transistor computers, this computer was smaller, more affordable, and faster. Furthermore, IC technology not only advanced general-purpose computer but also laid the foundation for the emergence and development of microcomputers and personal computers (as shown in Fig. 2.6).

The modern computer era represents a significant leap in hardware and software technologies, driving the advancement of various computing systems from microcomputers to supercomputers, as shown in Fig 2.7. In 1971, Intel introduced the first microprocessor, marking the dawn of the microcomputer age. This innovation drastically reduced the cost and size of computers while enhancing processing power. As a result, microcomputers surpassed the capabilities of earlier mainframes, enabling widespread adoption in homes and offices. Meanwhile, large-scale integration (LSI) and very large-scale integration (VLSI) technologies further propelled classical

Fig. 2.5 The 1956 Nobel Prize in Physics laureates (John Bardeen, Walter Brattain, and William Shockley). They were awarded for their groundbreaking invention of the transistor

Fig. 2.6 Fathers of the integrated circuit (Jack Kilby and Robert Noyce)

computing. By enabling millions of transistors to be integrated onto a single chip, these advancements significantly enhanced the overall speed and performance. The

Fig. 2.7 Different types of modern computers

combination of microprocessors with LSI and VLSI technologies led to the emergence of personal computers (PCs). In addition to microcomputers, supercomputers also advanced significantly. As their physical size decreased, their processing speeds reached billions of calculations per second, meeting the demands of increasingly complex and large-scale tasks. Building on these hardware advancements, software developments such as database management, network management, and object-oriented programming significantly expanded the scope of computer applications, enhancing both functionality and usability.

Thanks to the contributions of generations of researchers, classical computing systems have grown progressively more complex and versatile. Today, their applications extend across a broad spectrum of fields, from scientific research to everyday tasks and entertainment, profoundly influencing and transforming daily life in ways that were once unimaginable.

2.1.3 Quantum Computing

2.1.3.1 Quantum Mechanics

Democritus, an Ancient Greek pre-Socratic philosopher, is best known for formulating an atomic theory of the universe. His ideas sparked human curiosity about the microscopic world, particularly at atomic and subatomic scales. However, as scientists began to explore these scales, they encountered phenomena that classical physics could not explain (as shown in Fig. 2.8).

To address the inconsistencies that classical physics could not explain, quantum mechanics emerged as a theoretical framework. Initially, this development stemmed from efforts to resolve issues like Max Planck's solution to the black-body radiation problem in 1900, as well as Albert Einstein's photoelectric effect in 1905. These

Fig. 2.8 Cheerful Democritus painting by Charles-Antoine Coypel

early investigations into microscopic phenomena laid the groundwork for future exploration and are collectively referred to as the "old quantum theory."

The full development of quantum mechanics occurred in the mid-1920s, through the work of physicists such as Niels Bohr, Erwin Schrödinger, Werner Heisenberg, Max Born, and Paul Dirac. The modern theory is built upon sophisticated mathematical formalisms, including the well-known wave function, which provides information in the form of probability amplitudes about the potential measurements of a particle's energy, momentum, and other physical properties.

Research in this field has continued to flourish, garnering significant recognition, including Nobel Prizes. In 2022, physicists Alain Aspect, John Clauser, and Anton Zeilinger were awarded the Nobel Prize in Physics for their groundbreaking experiments on quantum entanglement—a phenomenon where two particles remain interconnected and share information regardless of distance, challenging conventional notions of communication (as shown in Fig. 2.9).

2.1.3.2 Quantum Computing

The concept of quantum computing emerged from the principles of quantum mechanics, revolutionizing the way computations could be performed. The idea was first introduced by physicist Richard Feynman in 1982 [53], who observed that classical computers were inefficient at simulating quantum systems due to their inherent limitations. He proposed that a new type of computer, operating on quantum mechanics, could be more effective for simulating such systems. In 1985,

Fig. 2.9 2022 Nobel Prize in Physics Laureates (John Clauser, Anton Zeilinger, and Alain Aspect)

David Deutsch[3] further developed the idea by formalizing the concept of a universal quantum computer, a machine capable of solving a wide range of computational problems by exploiting quantum phenomena.

Quantum computing is significantly different from classical computing. While classical computers process information in bits, which can be either 0 or 1, quantum computers utilize quantum bits, also known as *qubits*. Unlike bits, qubits can exist in multiple states simultaneously due to the phenomenon of *superposition*. Another key phenomenon in quantum computing is *entanglement*, where qubits become correlated in such a way that the state of one qubit is directly linked to the state of another, even when they are physically separated. This unique property enables quantum computers to perform specific computations that would be impossible or highly impractical for classical systems.

The potential of quantum computing has been demonstrated through several breakthrough algorithms. For example, Shor's algorithm provides a polynomial-time solution to integer factorization, which has profound implications for cryptanalysis. Similarly, Grover's algorithm offers a quadratic speedup for searching unsorted databases, highlighting the ability of quantum computers to solve optimization and search problems more efficiently than classical computers. These advancements emphasize the promise of quantum computing to surpass classical systems in specialized domain, such as cryptanalysis, material science, and optimization.

Neven's law states that quantum computers are improving at a "doubly exponential" rate. If it holds, quantum supremacy is around the corner. This trend can

[3] https://plato.stanford.edu/entries/qt-quantcomp/

Fig. 2.10 IBM quantum computer installed at Shin-Kawasaki

be seen clearly from quantum roadmap proposed by IBM.[4] As quantum computing continues to advance, we are witnessing growing investment and development from both academic institutions and technology companies. As shown in Fig. 2.10, an IBM quantum computer is installed at Shin-Kawasaki for the University of Tokyo, underscoring the expanding presence and capabilities of quantum computing in real-world applications.

2.1.4 *Quantum Arithmetic*

Quantum arithmetic is a fundamental building block of quantum computing, involving the design of efficient arithmetic modules using basic quantum operations. These modules include, but are not limited to, addition, subtraction, multiplication,

[4] ibm.com/roadmaps/quantum/

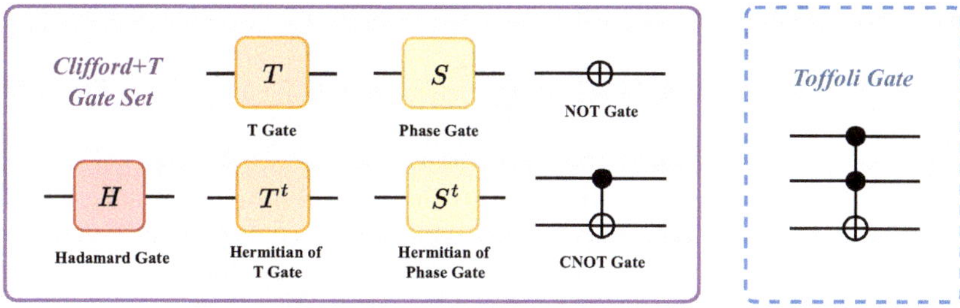

Fig. 2.11 Clifford+T gate set

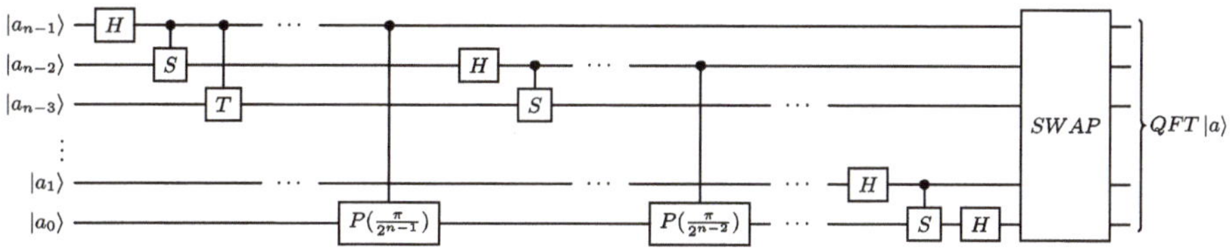

Fig. 2.12 Quantum Fourier transformation to a n-qubit quantum state $|a\rangle := |a_{n-1}a_{n-2}\cdots a_1a_0\rangle$

division, and modular exponentiation.[5] These quantum arithmetic modules play a vital role in many important quantum algorithms, such as Shor's Algorithm [193].

In the field of quantum computing, one main design method utilizes the Clifford+T gate set. Circuits designed with this method can leverage advanced fault-tolerant architectures based on quantum error-correcting codes (QECC), such as the surface code, thereby alleviating major challenges like the fragility of quantum states, short coherence times, and vulnerability to external noise. In this book, to standardize and compare various quantum designs, we discuss the T gate cost primarily at the level of the Toffoli gate. This method implements quantum circuits using Clifford gates (including CNOT, Hadamard, and phase gates) combined with T gates, which are detailed in Fig. 2.11. Although Clifford gates are relatively simple to implement and have lower error rates, T gates are complex and costly, making the optimization of T gate usage crucial for designing efficient quantum circuits.

Another mainstream arithmetic framework is based on quantum Fourier transformation (QFT). Relying on the principles of classical Fourier transform, QFT corresponds to the quantum counterpart that maps binary basis to Fourier basis [150]. QFT circuit has a fixed structure as is shown in Fig. 2.12, where $P(\phi) := |0\rangle\langle 0| + e^{i\phi}|1\rangle\langle 1|$ denotes the phase gate that rotates the quantum state around Z-axis in the block sphere with an angle of $\phi \in [0, 2\pi)$.

Comprehensive reviews of QFT-based arithmetic frameworks can be found in [9, 177]. QFT-based quantum arithmetic circuit typically begins with a QFT

[5] For a comprehensive overview of previous work on each quantum arithmetic operation, please refer to Sect. 3.6.

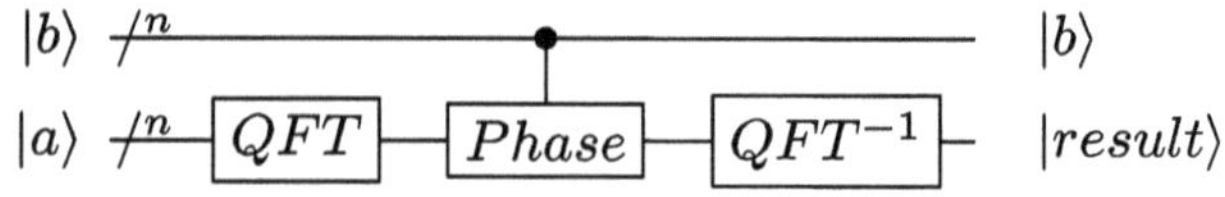

Fig. 2.13 Architecture of QFT-based arithmetic circuits

block converting the input state into the frequency domain, performing arithmetic operations using controlled phase gates, and then recovering the state to the original domain through inverse Fourier transform. This approach leverages efficient operations in the Fourier basis to simplify arithmetic operations, particularly in scenarios requiring multiplication and exponentiation. Figure 2.13 shows an illustrative diagram of the schematic of QFT-based quantum arithmetic circuit.

Quantum arithmetic circuits hold immense potential for accelerating large-scale quantum algorithms. For instance, in Shor's algorithm, efficient quantum arithmetic circuits can significantly reduce computational complexity and speed up the factorization process. Similarly, the efficiency of quantum arithmetic circuits is directly related to the overall performance of the algorithm.

2.1.5 *Number Formats*

When designing quantum arithmetic circuits, it is essential to consider number formatting. In this chapter, we focus on three main number formats for arithmetic designs as follows.

2.1.5.1 Integer

Integers are whole numbers without any fractional components and can be represented in various sizes, such as the commonly used 8-bit, 32-bit, or 64-bit formats. In both classical and quantum computing, integer operations are simple and efficient, making them widely applicable for fundamental tasks such as indexing, counting, and modular arithmetic. As a result, most arithmetic circuit designs in quantum computing focus on integer operations.

2.1.5.2 Fixed Point

Fixed-point numbers have a whole part and a fractional part, separated by a decimal point. This format allows for storing values with a fixed range and precision. Fixed-point arithmetic involves operations similar to those in integer arithmetic, but with the consideration of the position of the decimal point.

Compared to integer formats, fixed-point numbers have received less attention in the quantum computing field. Typically, the designs for quantum fixed-point arithmetic are very similar to those for integer arithmetic circuits with the main

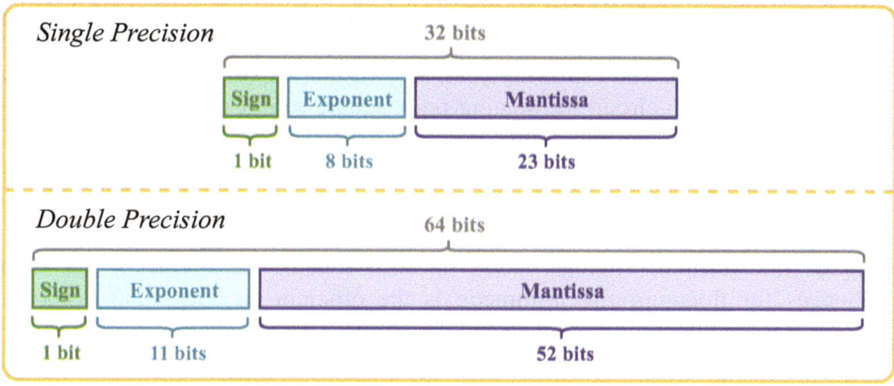

Fig. 2.14 IEEE-754 floating-point standard

difference lying in the representation block for fixed-point numbers. For example, several fixed-point quantum arithmetic designs proposed by Zhang [248] demonstrate structures very similar to corresponding integer operations.

2.1.5.3 Floating Point

Floating-point numbers each contain sign, exponent, and mantissa components. The representation of floating-point numbers offers the advantage of dynamic precision, resulting in a much larger range of values. Unlike fixed-point numbers, the precision of floating-point values can be dynamic and can be changed by manipulating the exponent and mantissa bits, while still being stored in fixed lengths. This property is crucial as it allows industry experts to design standardized number formats for computing, such as IEEE-754 (Fig. 2.14).

In the realm of quantum computing, floating-point arithmetic circuits are often extensions of integer designs, primarily involving specialized modules for floating-point representation and processing. Here, we present a comprehensive review of prominent research on quantum floating-point arithmetic designs.

In 2013, Nguyen et al. [147] introduced a space-efficient design for a reversible floating-point adder based on the IEEE-754 single-precision floating-point format, which is particularly well-suited for binary quantum computation. The proposed adder addresses the lack of existing designs for reversible floating-point adders at that time, providing a crucial advancement in the field. Following this, Jain et al. [102] proposed designs for reversible floating-point Ripple-Carry addition, multiplication, and division, all based on integer arithmetic blocks. Their work also utilizes the IEEE-754 standard for floating-point representation.

In 2018, Häner et al. [100] introduced quantum circuits for floating-point addition and multiplication using two distinct methodologies: automated circuit generation from classical Verilog implementations and manual circuit generation with optimization. In 2021, Gayathri et al. [179] proposed T-count and T-depth optimized quantum circuits for floating-point addition and multiplication using the IEEE-754 format. The following year, they further expanded their research by presenting a quantum circuit for floating-point square root operations based on the classical Babylonian algorithm [178].

In 2022, Zhao et al. [250] proposed an enhanced quantum floating-point adder based on the IEEE 754 standard. They categorized quantum floating-point adders into four cases: normal, subnormal, mixed, and special cases, and subsequently designed the quantum modules based on each specific case. One of the main challenges for floating-point arithmetic is the efficient encoding and decoding between classical floating-point data and the corresponding quantum states.

In 2022, Seidel et al. [188] presented an efficient design of unsigned integer arithmetic based on the fact that these evaluations can be written as semi-boolean polynomials. Intriguingly, the authors extended the framework to floating-point arithmetic, where non-integers are encoded such that the mantissa is quantum and the exponent is a classically known number, with the bit shift operations that map the non-integer coefficients to the corresponding unsigned integers to suit the semi-boolean polynomials. For the circuit design, QFT and its inverse are utilized as subroutines to construct the bit shift operators and reverse the data to the exact floating points before measuring the outcomes. Combining the bit shift procedure and the in-place operations, this design method provides a speedup by the parallelization of the gate executions over semi-boolean polynomials.

In brief, both quantum fixed-point and floating-point arithmetic designs heavily rely on their respective quantum integer designs. Therefore, for conciseness, we will primarily focus on integer number formats in the following chapters.

2.2 Quantum Architectures

In this book, we also consider various quantum hardware architectures, which are crucial for verifying the functionality of our designs. Currently, there are two main quantum computing architectures, as detailed below.

2.2.1 Noisy Intermediate-Scale Quantum Computing

The first type of quantum architecture discussed in this book is NISQ computing, a concept introduced by John Preskill [167]. This architecture is characterized by the noise generated during each quantum operation, which imposes significant limitations on circuit depth and overall computational capacity [12].

NISQ architectures can be realized using various technologies, including trapped-ion quantum devices, superconducting quantum devices, and photonic quantum devices. These "intermediate-scale" quantum processors typically contain up to 1,000 qubits. However, they remain insufficient for fault-tolerant quantum computing or achieving quantum advantage. A major challenge lies in their lack of error correction capabilities, which significantly limits their performance due to the noise inherent in each operation. This makes NISQ architectures highly susceptible to noise and decoherence.

Most designs for NISQ architectures utilize a quantum-classical hybrid approach, where complex computational tasks are partially processed by quantum computers, with the remaining tasks processed by classical devices. Despite the limited number of qubits, recent advancements have demonstrated NISQ devices are able to perform tasks that exceed the capabilities of classical computers. However, these devices are highly prone to errors due to noise and decoherence, which limits their feasibility significantly.

2.2.2 Fault-Tolerant Quantum Computing

In classical computing, fault tolerance is a fundamental concept where redundancy and error correction ensure the reliability of systems. In quantum computing, it is even more crucial because quantum states are much more delicate and highly susceptible to errors. To address the limitations of NISQ architectures, the field of quantum computing is shifting toward fault-tolerant quantum computing. This approach aims to construct quantum devices that can correct errors and perform computations with high accuracy.

So far, fault-tolerant quantum computing is in its early stages and faces several significant challenges, as outlined below:

- **Quantity of Qubits.** Current state-of-the-art devices typically have only a few hundred qubits, while fault-tolerant architectures need millions or even billions.
- **Quality of Qubits.** It is also crucial to maintain high-quality qubits that can preserve their quantum states over extended periods.
- **Error Susceptibility.** Existing qubits are highly susceptible to errors from noise and decoherence, which limits their effectiveness in error correction.

To achieve fault-tolerant quantum computing, various techniques have been proposed to overcome noise in quantum devices. For example, one of the most important techniques is the surface code [55], which arranges qubits into two-dimensional lattices that define an error correction code. Besides, recent breakthroughs by IBM [24] and Google [2] mark substantial progress toward scalable, fault-tolerant quantum computing. For instance, Bravyi et al. [24] introduce an end-to-end quantum error correction protocol using low-density parity-check codes. This approach encodes logical qubits using fewer physical qubits compared to the surface code and achieves an error threshold of 0.7%. Similarly, Google's new quantum chip, Willow [2], demonstrates quantum error correction below the surface code threshold. Willow achieves a logical error rate of 0.143% per error correction cycle, representing a significant milestone in fault-tolerant quantum computation.

Achieving fault-tolerant quantum computing opens the door to impactful applications, such as Shor's algorithm, which efficiently factors large numbers. It could potentially undermine the security of many public-key cryptography systems currently utilized in the real world. Additionally, fault-tolerant architecture is also crucial for attaining quantum advantage in applications including optimization and finance. In these domains, the speedups achieved are often quadratic at best, meaning that the quantum runtime scales as the square root of the classical runtime. Consequently, the advantages of quantum computing primarily manifest for deep circuits, which are challenging to execute without effective error correction.

2.3 Applications

Quantum arithmetic plays a crucial role in algorithm implementations. A small improvement in arithmetic design can lead to significant reductions in algorithmic resource cost and computational complexity. Besides modular exponentiation in Shor's factorization algorithm, quantum arithmetic circuits have wide applications in quantum finance, quantum machine learning, quantum linear systems solvers, and accessing block-encoding oracles. Here, we briefly summarize the potential applications by elaborating an exemplar in each of the four areas. Moreover, we summarize a list of the algorithms and arithmetic kernels in Table 2.1:

- **Quantum Financial Modeling—Calculating Payoff Functions.**
  Quantum finance [50, 94, 157], also known as quantum algorithms for computational finance [171–173], involves using quantum algorithms to solve classical problems in the financial world with potential advantages. Derivative option pricing is crucial for maximizing returns based on the future price of an underlying asset, such as stocks, currencies, or goods, and a strike price agreed upon in a contract between a buyer and seller. The most typical payoff function is the European call option, which gives the buyer the right, but not the obligation, to exercise the option at the strike price by the expiration date [19, 141]. The buyer's return is calculated by a nonlinear function that subtracts the strike price from the stock price if the stock price is higher than the strike price, or outputs zero if the stock price is lower.
  In the quantum world, the dynamics of asset prices are characterized by stochastic processes, which can be predicted using quantum amplitude estimation [199], quantum Monte Carlo [213], quantum signal processing [200], and other methods. However, most algorithms require oracles or subroutines based on fault-tolerant error-corrected computations, resulting in high quantum resource costs, as discussed by Chakrabarti [29]. Thus, it remains challenging to develop efficient and cost-effective arithmetic designs for calculating payoff functions in derivative pricing. In 2023, Lim et al. proposed spatially and temporally efficient circuits using quantum adders and subtractors with elementary Clifford gates and Toffoli gates [126]. They systematically compared the performance metrics

Table 2.1 An overview of prominent applications of quantum algorithms with their quantum arithmetic designs

Year	Application	Arithmetic kernel	Algorithmic design
1996	Cryptanalysis	Modular exponentiation	Shor's factorization algorithm [193]
2009	Linear systems solver	Modular exponentiation	HHL with quantum phase estimation [88]
2021	Linear systems solver	Modular exponentiation	NISQ-HHL with iterative QPE [237]
2021	Financial modeling	Clifford+T and QFT adder	Derivatives pricing using various arithmetic [29]
2022	Machine learning	Clifford+T adder	Implementation of ReLU with SWAP gates [252]
2022	Linear algebra	Clifford+T and QFT adder	Block encoding of classical data [34]
2022	Linear algebra	Clifford+T and QFT adder	Block encoding of sparse matrices [27]
2023	Financical modeling	Clifford+T Subtractor	Calculation of payoff functions [126]
2023	Linear systems solver	QFT-based adder	Shift operators for Hadamard test [99]
2023	Linear algebra	Clifford+T and QFT adder	Block encoding of structured matrices [202]
2024	Machine learning	Clifford+T adder	Constant T-depth ReLU [252]
2024	Linear algebra	Clifford+T adder	Block encoding of banded matrices [79]

of various quantum adders and used the two's complement method to realize the subtractor. By optimally choosing the best arithmetic kernels, they greatly reduced the resource cost for payoff function calculations and provided efficient options for hardware execution.

- **Quantum Machine Learning—Implementing Activation Functions.** Quantum machine learning has been a popular topic recently [18, 28, 98]. Among all the models, the most well-known is the quantum neural network, which takes quantum states as input to the first layer of nodes and evolves to another quantum state after passing through trainable parameterized hidden layers, with information readout via measurements [1, 77, 110, 146]. Problem specifications are encoded through the customization of the loss function. Optimization techniques are used to find the optimal parameters, converging to the solutions that correspond to the minimum of the loss function. The transformation from input states to the output states through neural networks is characterized by linear or nonlinear activation functions [137]. One of the most common and effective activation functions is the ReLU. ReLU is a nonlinear function that outputs the value if it is larger than zero and outputs zero for negative values.

In various models of quantum neural networks, the arithmetic design of ReLU usually involves high implementation complexity. In 2022, Sajadimanesh et al. proposed an efficient implementation of the ReLU activation function for an n-qubit state using only one depth of controlled SWAP gates and SWAP gates, making it suitable for NISQ executions [182]. However, this design requires n auxiliary qubits, which doubles the qubit count. Recently, Zi et al. proposed an efficient circuit design for the implementation of the ReLU with constant T-depth [252]. They leveraged the quantum fan-out gate and efficient Toffoli gate decomposition to calculate the activation function. They also discussed the implementation of other activation functions using the quantum lookup table, bridging the gap between algorithms and real-world implementations.

- **Quantum Linear Systems Solver—Realizing Shift Operators.**
Another quantum algorithm that can potentially outperform its classical counterpart is the linear systems solver. The most famous quantum algorithm for solving linear systems of equations is the HHL algorithm, proposed by Harrow, Hassidim, and Lloyd in 2009 [88]. It utilizes quantum phase estimation (QPE) as a subroutine to achieve a quadratic speedup, but it requires a large amount of resources. In 2021, Yalovetzky improved it by using the semiclassical Fourier transformation for iterative phase estimation in the HHL algorithm [237]. In 2019, Huang et al. proposed a near-term version of a linear systems solver with a proof of convergence guarantee [97]. This algorithm requires Hadamard tests and convex optimizations to efficiently obtain a classical solution. However, the Hadamard test requires efficient and cost-effective implementation of problem-aware unitaries, limiting the capability of general linear systems solvers with random problem matrices.

In solving differential equations, finite difference methods are used to discretize the system into linear systems of equations with banded circulant matrices. This matrix decomposes into a linear combination with powers of parameter shift operators, which have higher complexity in terms of circuit implementation using conventional Clifford+T designs due to the complicated decomposition process [234]. Given an n-qubit quantum state $|a\rangle = \sum_{i=0}^{N-1} a_i |i\rangle$ and N=2^n, a parameter shift operator is a periodic mapping that shifts computational basis $|i\rangle$ to $|i+1\rangle$ for any $i \in [N]$ and shifts the basis $|N-1\rangle$ back to $|0\rangle$. Inspired by Draper's quantum adder [45], Koch proposed an efficient scheme for parameter shifting based on QFT and single-qubit rotations [117]. This approach avoids the difficulties in Clifford+T design but requires an additional QFT block and its inverse. Based on this design, Huang et al. implemented the parameter shift operators in a Hadamard test using a QFT block and controlled phase gates. They proposed a hybrid quantum-classical algorithm with an improved performance guarantee for solving banded linear systems of equations [99].

Although QFT and its inverse are still resource-intensive, we can consider extending the framework of quantum linear systems solvers, previously designed solely for NISQ, to a stage where quantum devices can perform Fourier transform correctly and efficiently [99]. This insight can lead to discussions on early fault-tolerant quantum algorithms with randomized Fourier arithmetic, bridging the gap between current hardware platforms and future linear-systems-solver-based applications, including partial differential solvers, matrix inversion, and optimizations [112, 120, 239].

- **Quantum Linear Algebra—Accessing Block-Encoding Oracles.**
For most fault-tolerant algorithms, it is essential to equip the data with block-encoding matrices [202]. Given an n-qubit non-unitary operator A with dimension N=2^n, an m+n qubit unitary U is a (α, m, ϵ)-block-encoding of A if $||A - \alpha(\langle 0|^{\otimes m} \otimes I_N)U(I_N \otimes |0\rangle^{\otimes m})||_2 \leq \epsilon$ [66]. Based on the block structure, quantum linear and nonlinear algebra can be efficiently implemented using quantum eigenvalue transformation [134] or quantum singular value transformation [66, 76]. Although the framework seems promising, designing efficient circuits for constructing block-encoding matrices for arbitrarily given matrices remains challenging [34].
In 2022, Camps et al. proposed an exact circuit construction for well-structured sparse matrices using Clifford+T-based modular adders and QFT-based controlled phase gates [27]. Later in 2023, Snderhauf proposed an approximate implementation of a block-encoding oracle with in-place additions composed of Clifford+T gates and controlled rotation gates [202]. Furthermore, Guseynov et al. studied the Hamiltonian needed for simulating partial differential equations and proposed an explicit gate construction of block-encoding matrices [79].

2.4 Chapter Summary

This chapter provides a comprehensive review of previous work in quantum arithmetic, beginning with a brief background. We then explore relevant quantum architectures for mapping our designs to quantum hardware, including NISQ architecture and fault-tolerant architecture. Moreover, we also discuss the applications of quantum arithmetic.

Quantum arithmetic circuits have garnered widespread attention, not only in academia but also in the industry. Leading technology companies such as Google and IBM have invested significant resources and research interest in this field. These companies aim to advance the practical application and performance of quantum computers through the study and application of quantum arithmetic circuits. As quantum computing technology advances, the demand for quantum arithmetic circuits will continue to grow. Their broad application prospects in accelerating quantum algorithms, quantum communication, and quantum cryptanalysis herald the arrival of the quantum computing era. Research on quantum arithmetic circuits is not only theoretically significant but also poised to bring substantial technological breakthroughs and innovative applications.

> **Key Takeaways**

1. Quantum arithmetic circuits are essential for enhancing the performance and efficiency of quantum algorithms and applications.
2. Both NISQ and fault-tolerant architectures play crucial roles in the implementation of quantum arithmetic designs.
3. The growing demand for quantum arithmetic circuits aligns with the broader adoption of quantum computing technology.
4. Leading companies like Google and IBM are actively advancing the field through research and practical applications.
5. Continued research in this field promises significant theoretical insights and practical innovations.

Chapter 3
Efficient Quantum Arithmetic Designs

There's a way to do it better – find it.

–Thomas Edison

This chapter focuses on the core contribution of this book: the design and optimization of quantum arithmetic circuits. These circuits, including addition, subtraction, modular addition, and division, serve as foundational building blocks in many quantum algorithms. Among them, quantum addition is of particular importance, as it forms the basis for implementing other arithmetic operations. Consequently, we explore three different design perspectives to optimize quantum addition circuits. Building on quantum addition, we also discuss the realization of quantum subtraction and multiplication. Subtraction can be efficiently constructed using modified adder structures, and Sect. 3.6.3 presents two commonly used quantum subtraction designs derived from our proposed adder modules. For multiplication, which is already a well-established domain in the literature, we do not introduce new designs here, but rather provide a detailed review of existing quantum multiplier architectures in Sect. 3.6.4. Division, particularly in the existing non-restoring quantum division, remains less explored due to its nonstandard calculation. In this chapter, we propose an efficient quantum non-restoring divider architecture and analyze its performance under various design choices. To synthesize the contributions and contextualize their significance, Sect. 3.6 concludes the chapter with a comprehensive overview of recent advances in quantum arithmetic—including both prior works and our own designs.[1]

[1] This chapter includes our papers [224, 226, 229–232].

S. Wang, A. Chattopadhyay, *Quantum Arithmetic*, Computer Architecture and Design Methodologies, https://doi.org/10.1007/978-981-95-6039-4_3

3.1 Addition: Quantum Higher Radix Adder

To perform the addition of two n-bit numbers, the quantum adder proposed in design 1 utilizes $O(n) - O(\frac{n}{r})$ qubits and $O(n) + O(\frac{n}{r})$ T gates. This design achieves outstanding performance in T-depth, specifically with $O(r) + O(\log \frac{n}{r})$, where r denotes the radix.

3.1.1 Introduction

Despite the comprehensive analysis of adders in classical computing, this specific area has not been thoroughly explored in the quantum computing domain. The initial quantum adder introduced the quantum ripple carry adder (RCA) [40, 218] into quantum world, which, unfortunately, suffers from linearly increasing T-depth. Subsequent quantum carry look-ahead adders (CLA), such as Draper's logarithmic adder [46], have achieved efficiency gains with logarithmic T-depth. Surprisingly, although classical adders have explored the benefits of increasing the radix of CLA, this higher radix strategy has not been thoroughly explored in quantum CLAs.

This study seeks to explore the potential of the higher radix strategy in enhancing quantum addition circuits. To address challenges like high fan-out in quantum adders, we introduce an innovative quantum higher radix adder. This quantum circuit comprises two significant parts: The higher radix part utilizes Gidney's Logical-And [62] and Selinger's Multi-control Toffoli construction method [189] to create a general quantum higher radix layer. In Manchester Carry Chain (MCC) section, we use the Brent-Kung structure as the carry path and Gidney's Adder [62] as the sum path.

The contribution of this work lies in the improved efficiency of quantum circuits across various scales by using this novel quantum higher radix adder. Besides, we also aim to provide deeper insights into the potential of multi-control Toffoli gates and the higher radix strategy in enhancing quantum arithmetic circuits. Moreover, we illustrate that quantum adders can be categorized based on Qubit Count (QC), T-count, and T-depth, aligning with Harris' classification of classical adders [86]. All relevant source codes are available on GitHub.[2]

3.1.2 Methodology

Here, we firstly introduce C_nNOT, a crucial component of the proposed quantum higher radix adder. Our specific method involves the following key steps: Firstly, we

[2] https://github.com/Siyi-06/Quantum_higher_radix_adder

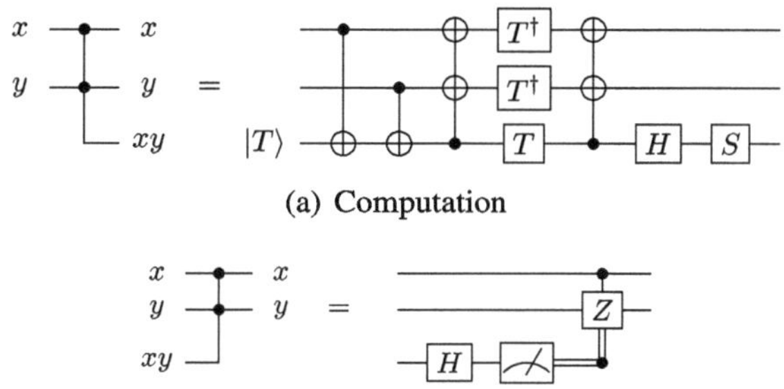

(a) Computation

(b) Uncomputation

Fig. 3.1 Gidney's Logical-And structure

outline the construction of the higher radix layer using the C_nNOT gate. Then, after conducting an analysis of five carry-propagate structures and two sum structures, the Brent-Kung structure and Gidney's RCA are selected as the Carry path and Sum path, respectively. At the end, we provide a detailed explanation of how to construct the overall quantum higher radix circuit.

3.1.2.1 Basic Component: C_nNot

C_nNOT is a fundamental structure in the proposed quantum higher radix adder. To build an effective C_nNOT gate, we employ and enhance Gidney's Logical-And structure illustrated in Figs. 3.1a and b.

In Gidney's work [62], the first expression in Eq. (3.1) defines the special state $|T\rangle$ within the Logical-And structure in Fig. 3.1a. Following this formula, we can apply a Hadamard gate followed by a T gate on an ancilla in the $|0\rangle$ state to obtain $|T\rangle$. Interestingly, we found that it is feasible to use an ancilla in the $|1\rangle$ state instead of $|0\rangle$ to construct this structure, as indicated in the second part of Eq. (3.1). Obviously, quantum operations like quantum NOT can be performed before the Logical-And structure, contributing to a reduction in the number of qubits required for the proposed higher radix quantum adder.

$$|T\rangle = \begin{cases} \frac{1}{\sqrt{2}}(|0\rangle + e^{i\pi/4}|1\rangle) & \text{if } ancilla = |0\rangle; \\ \frac{1}{\sqrt{2}}(|0\rangle - e^{i\pi/4}|1\rangle) & \text{if } ancilla = |1\rangle. \end{cases} \tag{3.1}$$

After expanding the applicability of Gidney's Logical-And, we proceed to demonstrate the specific structure and decomposition method of the proposed multi-control Toffoli gate. Table 3.1 demonstrates that as the number of control qubits increases, the C_nNOT gate effectively reduces the average T-count, T-depth, and QC per control qubit. Hence, we employ multi-control Toffolis in this study to minimize circuit costs.

Selinger proposed a general method in 2013 [189] for constructing C_nNOT gates based on Clifford gates and T gates. In this book, we optimize Selinger's

Table 3.1 Cost of C_nNOT

Structure	T-count	T-depth	Qubit Count
C_2NOT	TC	TD	Q
C_3NOT	$4+TC$	$2+TD$	$2+Q$
C_4NOT	$8+TC$	$3+TD$	$4+Q$
C_5NOT	$12+TC$	$3+TD$	$6+Q$
C_6NOT	$16+TC$	$4+TD$	$8+Q$
C_{n+2}NOT	$4n+TC$	$2+\lfloor \log n \rfloor + TD$	$2n+Q$

$TC = 7, TD = 3, Q = 3$

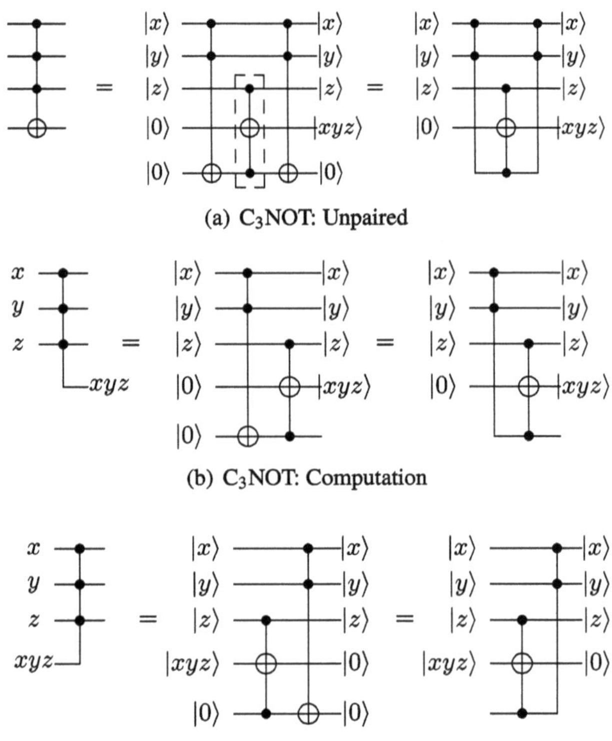

Fig. 3.2 Construction of multi-control Toffoli

(a) C_3NOT: Unpaired

(b) C_3NOT: Computation

(c) C_3NOT: Uncomputation

approach by incorporating ideas from the Logical-And structure [62] and another Toffoli decomposition method [5]. Figure 3.2 illustrates the detailed structure of the proposed multi-control Toffoli gate. Utilizing Logical-And structure, here we construct a C_3NOT gate as an example. Specifically, Fig. 3.2a demonstrates the design of an unpaired C_3NOT gate based on Clifford + T gates. For a pair of C_3NOT gates, Figs. 3.2b and c illustrate the computation and uncomputation circuits, respectively. In Fig. 3.2a, b, and c, the leftmost circuits demonstrate the original designs, the middle circuits demonstrate the decomposed circuits based on Toffoli gates, and the rightmost circuits decomposed circuits using the Logical-And structure.

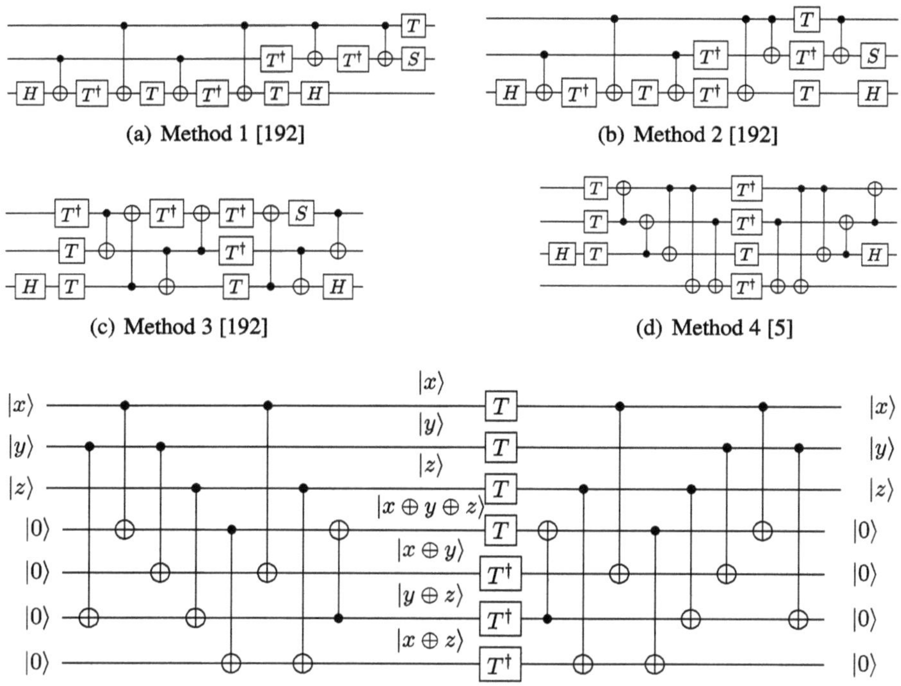

(a) Method 1 [192]

(b) Method 2 [192]

(c) Method 3 [192]

(d) Method 4 [5]

(e) Method 5 [192]

Fig. 3.3 Different methods to decompose unpaired Toffoli with Clifford + T gates [5, 189]

Specifically, employing computation and uncomputation structures of Logical-And to decompose the remaining Toffoli gates except the middle one effectively reduces the T-count and T-depth of the proposed multi-control Toffolis, enhancing the overall circuit efficiency.

Figure 3.3 presents five general decomposition methods for the middle unpaired Toffoli [5, 189]. According to Table 3.2, we find that all the decomposition methods use 7 T gates. Since the proposed higher radix adder has a high QC, in Study 1 we try to avoid introducing more ancilla qubits by avoiding decomposition methods with ancilla qubits. Specifically, method 5 provides a T-depth of 1 for the circuit resulting from the decomposition of a single unpaired Toffoli. However, it introduces four additional ancilla qubits, significantly increasing the number of qubits required for constructing a CLA. Therefore, we omit Method 5 from consideration. Similarly, Method 4 also involves introducing extra ancilla qubits, which makes it less favorable in our selection process. Among Methods 1, 2, and 3, Method 3 has the highest efficiency due to its minimal T-depth. Hence, we choose Method 3 to decompose all unpaired Toffolis in our work.

Table 3.2 Summary table of Toffoli decomposition

Decomposition	T-count	T-depth	Ancilla count
Method 1 (Fig. 3.3a)	7	6	0
Method 2 (Fig. 3.3b)	7	4	0
Method 3 (Fig. 3.3c)	7	3	0
Method 4 (Fig. 3.3d)	7	2	1
Method 5 (Fig. 3.3e)	7	1	4

3.1.2.2 Step 1: Higher Radix Layer

In this section, we introduce the higher radix strategy and implementation details for applying it to quantum circuits. Radix is the bit width of each CLA block in carry look-ahead adders. For convenience, we use r to represent radix. In the classical world, the higher radix adder has shown its advantages [80]. By increasing the radix, an adder with higher fan-in and fan-out is constructed, which can reduce the propagation time of propagates (p) as well as generates (g), thus improving the efficiency of classical adder.

We firstly take the radix 2 CLA as an example to calculate the sum of two binary numbers a and b. We need to get the carry c after 2-step calculations. In the first step, we calculate the p and g of each bit. For the ith bit, we use the formulas 1 and 2 to compute the corresponding p_i and g_i, respectively. In the second step, after computing the propagation of p and g to obtain the intermediate quantities P and G, we can calculate the carry c. Here, we assume that j is less than or equal to i. Equations (3.4), (3.5), and (3.6) describe how to propagate p and g from bit i to bit j. After using the formulas (3.7), we are able to obtain the final carry c. The standard notations of $\oplus$ for logical XOR, $\cdot$, $+$, $\circ$ for logical AND, logical OR, and propagation, respectively, are here.

$$p_i = a_i \oplus b_i \tag{3.2}$$

$$g_i = a_i \cdot b_i \tag{3.3}$$

$$(G_{0:0}, P_{0:0}) = (g_0, p_0) \tag{3.4}$$

$$(G_{0:i}, P_{0:i}) = (g_i, p_i) \circ (G_{0:i-1}, P_{0:i-1}) \tag{3.5}$$

$$(g_x, p_x) \circ (g_y, p_y) = (g_x + p_x \cdot g_y, p_x \cdot p_y) \tag{3.6}$$

$$c_i = g_i + p_i \cdot c_{i-1}. \tag{3.7}$$

As shown in Fig. 3.4, we use multi-control Toffoli gates to construct the quantum propagation structures with radix 2, 3, and 4, respectively. Similarly, the higher radix strategy also reduces the number of propagation layers of p and g, thereby reducing T-count required for the addition operations, and further effectively reducing the operation time. However, due to the extremely high cost of fan-out larger than 2, the quantum higher radix strategy proposed only focuses on increasing the fan-in of the adder.

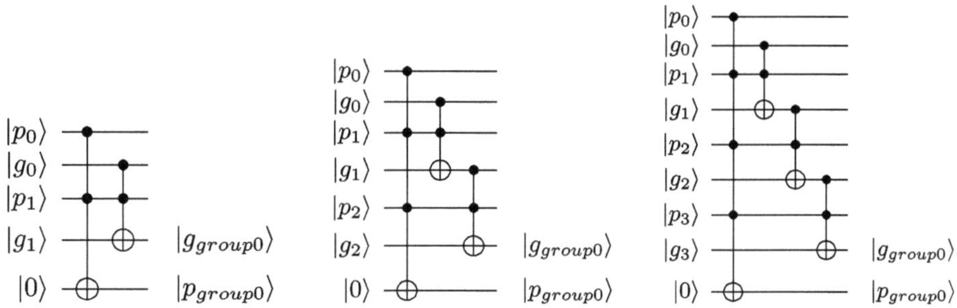

(a) Radix 2 (b) Radix 3 (c) Radix 4

Fig. 3.4 Quantum circuits for higher radix layer

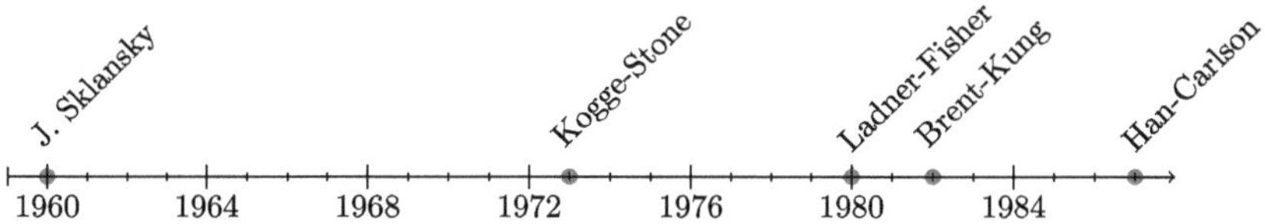

Fig. 3.5 Chronology of publication of carry-propagate structures

In this study, we try to construct a quantum higher radix adder, which means we need to construct the higher radix layer based on multi-control Toffolis for the propagation of p and g. Figures 3.4b and c show the specific circuits for the higher radix structure with radix 3 and radix 4 as respective examples.

3.1.2.3 Step 2: Carry Path

In this section, we describe the details of our carry path. Through the calculation in the previous section, we have already obtained p and g for each bit. In order to get the carries, we need to select a particular carry-propagate structure as the carry path to propagate p and g. As shown in Fig. 3.5, five different propagation structures are proposed in the literature, which are subsequently discussed. The details of these structures are shown in Fig. 3.6.

- **Sklansky.** In 1960, J. Sklansky [195] proposed a conditional CLA adder with high fan-out nodes and minimal depth. The structure of it is shown in Fig. 3.6a.
- **Kogge-Stone.** The Kogge-Stone structure [118] was published in 1973, which has a low depth but a high number of nodes. The structure is shown in Fig. 3.6b.
- **Ladner-Fisher.** As shown in Fig. 3.6a, the topology of Ladner-Fisher [121] is similar to the Sklanskly structure. Hence, it also has low depth but high fan-out nodes. However, there are some differences between these two structures in the application.

Fig. 3.6 Carry-propagate structures, where the green node represents one propagate operation

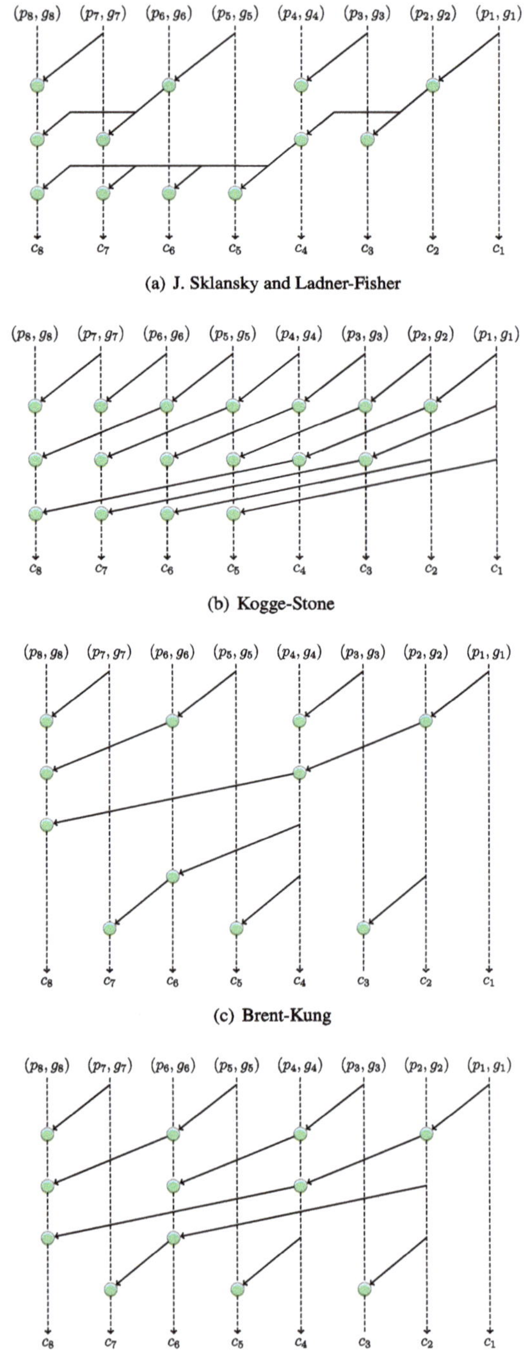

(a) J. Sklansky and Ladner-Fisher

(b) Kogge-Stone

(c) Brent-Kung

(d) Han-Carlson

- **Brent-Kung.** The Brent-Kung structure [25] is one of the most important propagation structures. Compared to other structures, this structure has a very small number of nodes as well as low fan-in and fan-out, despite having a large logic depth. Therefore, it is widely used in quantum CLA designs.
- **Han-Carlson.** In 1987, the Han-Carlson structure was first proposed [82]. In order to improve the overall efficiency of the propagation, it combines elements from both the Brent-Kung and Kogge-Stone structures.

The propagation operations are the main cost of the carry path. Furthermore, since qubit cannot be copied, the carry-propagate structures with fan-out larger than 2 introduce additional cost. Therefore, as shown in Fig. 3.6, the Brent-Kung structure, which has the smallest number of propagate operations and low fan-out, is selected as the carry path for the p and g propagation in our work.

3.1.2.4 Step 3: Sum Path

We then discuss the implementation details of the sum path. As described in the previous sections, the final sum can be calculated by feeding carries into the sum path. For the sum path, we can choose between CSA and RCA:

- **RCA.** The general structure of RCA is shown in Fig. 3.7a. Various quantum RCAs have been proposed so far. In this chapter, we choose the most efficient Gidney adder, which has the minimum T-count and T-depth, as our RCA structure.
- **CSA.** As shown in Fig. 3.7b, the Carry Select adder consists of two Ripple Carry adders and one select circuit.
  In the first part, we built two quantum RCAs with the same structure. The input carry bits of them are set to 0 and 1. Therefore, using these sub-circuits, we can obtain the sum when the inputs are 0 and 1, respectively. In both the classical and quantum worlds, this part can be computed in parallel with the carry path, thus effectively reducing the time cost.
  In the second part, we construct a select sub-circuit. Suppose we know that the real input carry is c, which can only be 0 or 1. Depending on c, the sum calculated by the corresponding RCA is then chosen as the final result. More specifically, when c is equal to 0, we choose the sum of the quantum RCA whose input carry is 0 as the final sum. Similarly, when $c = 1$, the final result is the sum of quantum RCA whose input carry is 1.
  It is worth noting that the quantum CSA contains expensive CSWAP gates, which can be decomposed by Clifford+T gates. As shown in Fig. 3.8, there are two decomposition methods. For Method 1 [5], the CSWAP gate is decomposed into a Clifford+T circuit with seven T gates and a T-depth of 4. If we use Method 2 (this is adopted from Quipper documentation (https://www.mathstat.dal.ca/~selinger/quipper/doc/Quipper-Libraries-GateDecompositions.html)), the CSWAP gate is decomposed into a circuit with seven T gates incurring the T-depth of 3.

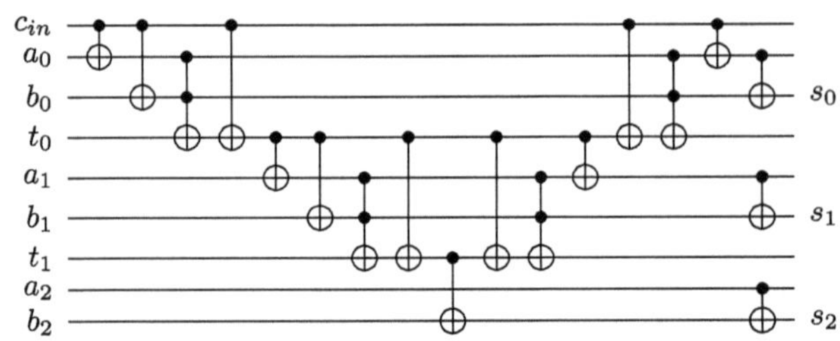

(a) RCA

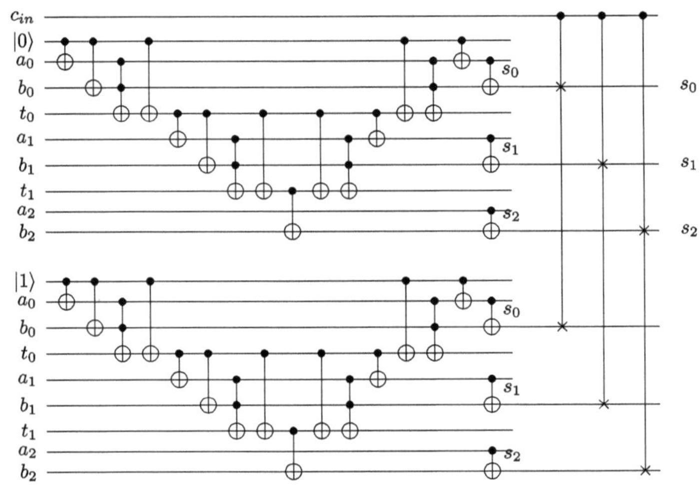

(b) CSA

Fig. 3.7 Sum paths

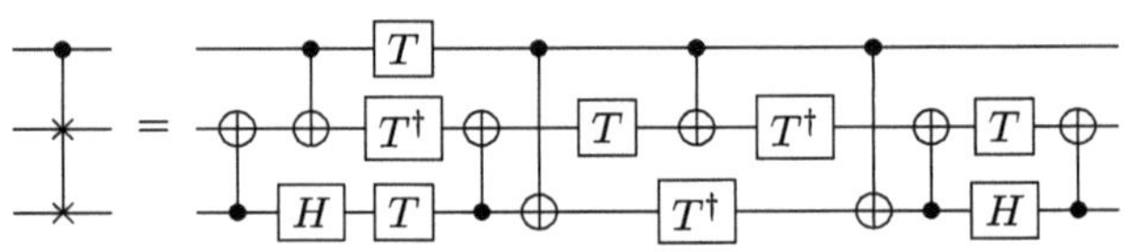

(a) Method 1

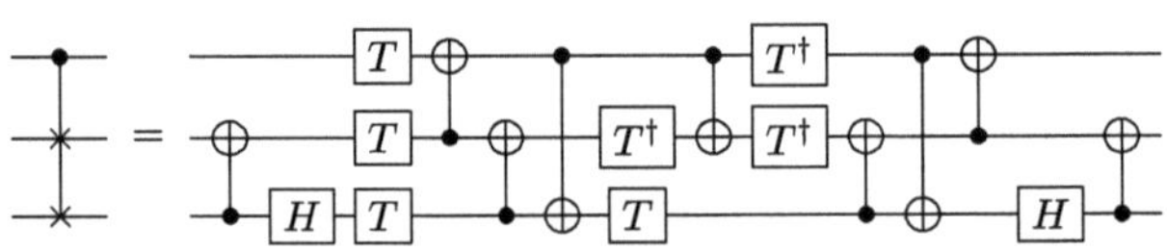

(b) Method 2

Fig. 3.8 Decomposition methods for quantum CSWAP gate

Table 3.3 Different structures of sum path ($r = n$). When calculating T-depth, we assume that Part 1 of the CSA has been completed when calculating the carry path. Therefore, for CSA1 and CSA2, T-depth equals to the T-depth of Part 2, which is the minimum T-depth of sum path for CSAs

Structure	T-count	T-depth	QC
CSA1	$11n - 4$	$4n$	$6n + 1$
CSA2	$11n - 4$	$3n$	$6n + 1$
RCA	$4n - 4$	n	$3n$

In this chapter, we use CSA1 to denote the CSA structure whose CSWAPs are decomposed by Method 1. Similarly, CSA2 represents the CSA whose CSWAPs are decomposed by Method 2.

After determining the design details of these sum paths, we performed a systematic analysis of their performance. As shown in Table 3.3, the RCA structure is always cheaper than the CSA structures in terms of T-count, T-depth, and QC. Therefore, we choose Gidney's RCA as our sum path.

3.1.2.5 Overall Structure of Quantum Higher Radix Adder

One Higher Radix Layer

In Study 1, we only use one higher radix layer. As shown in Fig. 3.9a (this diagram was inspired from https://web.stanford.edu/class/archive/ee/ee371/ee371.1066/lectures/lect_04.pdf), the classical higher radix adder applies the higher radix strategy to every layer of the Brent-Kung tree, which reduces the computation depth from $log2^n$ to log_r^n.

In this part, we explain why do we not follow the same example from the classical computing. This is due to the fact that if the strategy is used at every layer as shown in Fig. 3.9b, RCAs with large T-depth will be introduced in the sum path, which deprives our higher radix adder of the significant advantage of low T-depth according to Fig. 3.10. Therefore, we recommend using the higher radix strategy only for the first few layers of the quantum Brent-Kung tree.

Overall Structure

The structure of the proposed quantum higher radix circuit is shown in Fig. 3.11. It can be divided into seven stages:

- **Notations.** The binary bit width of the addends is denoted as n, and r denotes the value of the radix. The binary expansion of the number a is denoted as $a = a_{n-1}a_{n-2}\cdots a_0$, where a_{n-1} is the most significant bit and a_0 is the least significant bit. For circuit decomposition, paired Toffoli gates are decomposed

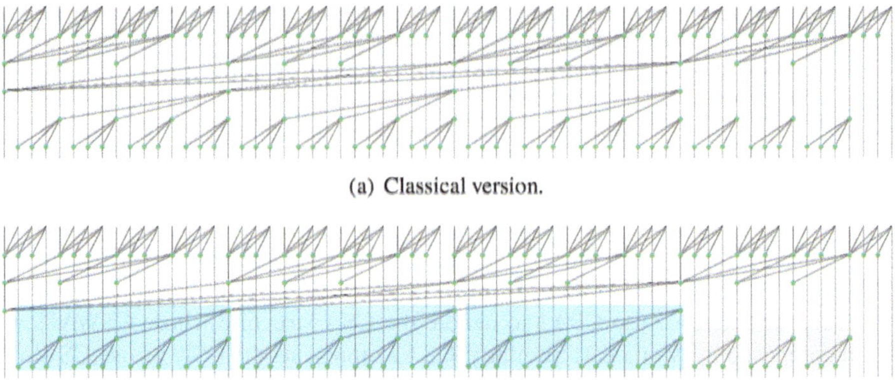

(a) Classical version.

(b) Quantum version.

Fig. 3.9 Brent-Kung structure with radix 4 (the small bounding box represents radix-4 RCA, and the large bounding box represents radix-16 RCA)

Fig. 3.10 The T-depth of the sum path with increasing layers for the radix-4 higher radix adder. The formula represented by this curve is $O(T - depth) = r^{layers}$

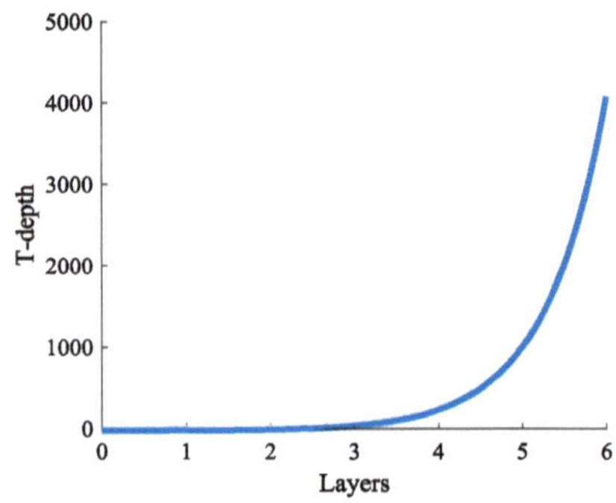

into Logical-And, while unpaired Toffoli gates are decomposed using Method 3 shown in Fig. 3.3. Thus, TC_3 is equal to 7, and TD_3 is equal to 3.

- **Step 1.** In this step, our task is to calculate the p and g. Since we do not need the most significant carry, no operation is performed on the most significant group. By taking a_i, b_i as control qubits and an ancilla with initial state $|0\rangle$ as the controlled qubit, we apply the CCNOT gate to compute g_i and then store the result in the corresponding ancilla. After that, we use a_i as the control and b_i as the controlled qubit to apply the CNOT gate. As a result, the corresponding p_i is stored in the corresponding b_i position. Some Toffoli gates are unpaired in the whole circuit, and we decompose those Toffoli gates by using Method 3.

  For convenience, α is introduced to denote the addend qubits in the most significant group. This step requires $TC_3 \cdot (n - \alpha)$ T-count, TD_3 T-depth, and extra ancilla qubits are $3 \cdot n - \alpha$.

$$\alpha = \begin{cases} r & \text{if } n \pmod{r} = 0; \\ n \pmod{r} & \text{otherwise.} \end{cases} \tag{3.8}$$

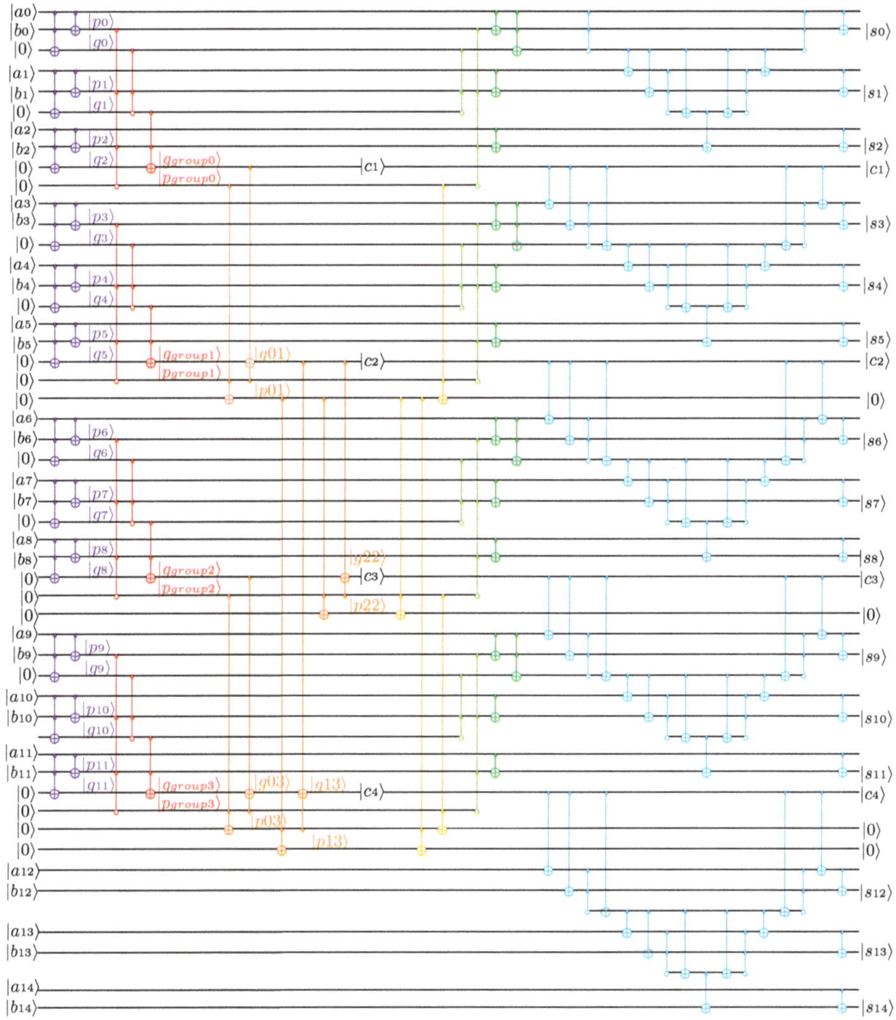

Fig. 3.11 Radix-3 addition circuit with two 15-bit addends a and b

- **Step 2.** In the second step, we group the initially obtained p and g by using the higher radix structure. Specifically, we construct the corresponding higher radix structure according to the method shown in Fig. 3.4 and then apply it to the corresponding g_i and p_i calculated in step 1 to obtain g_{group} and p_{group}. Since the controlled qubits of the last Toffoli will be used to store the carry later, no uncomputation is performed on it. Hence, the last Toffoli is always unpaired, and we only decompose it using Method 3, but decompose the rest of the Toffolis into Logical-And.

For convenience, β and ρ are introduced to represent a complex intermediate variable for constructing multi-control Toffolis and the number of groups divided, respectively. In step 2, the required T-count is $\rho \cdot [TC_3 + 4 \cdot (2 \cdot r - 3)]$, the T-depth is $TD_3 + r + \beta - 1$, and extra ancilla is $\rho \cdot (r - 1)$.

$$\beta = \begin{cases} 0 & \text{if } r \leq 2; \\ 2 + \lfloor \log(r-2) \rfloor & \text{otherwise} \end{cases} \tag{3.9}$$

$$\rho = \begin{cases} \frac{n}{r} - 1 & \text{if } n \pmod r = 0; \\ \lfloor \frac{n}{r} \rfloor & \text{otherwise} \end{cases}. \tag{3.10}$$

- **Step 3.** In step 3, we construct the Brent-Kung tree using the p_{group} and g_{group} processed by the higher radix structure to calculate the carry path. We tried using Logical-And here, but found no benefit. Hence, here all the Toffolis are decomposed by Method 3.
  In this step, the required T-count is $2 \cdot TC_3 \cdot [2 \cdot \rho - 1 - \omega(\rho) - \lfloor \log(\rho) \rfloor]$, the T-depth is $TD_3 \cdot (\lfloor \log(\rho) \rfloor + \lfloor \log \frac{\rho}{3} \rfloor + 2)$, and the number of extra ancilla qubits is $2 \cdot \rho - 1 - \omega(\rho) - \lfloor \log(\rho) \rfloor$.
- **Step 4.** In this step we uncompute the operation of calculating the intermediate p in step 3. We repeat the calculation of all Toffolis for the intermediate variable p in reverse order.
  In step 4, the required T-count is $TC_3 \cdot [2 \cdot \rho - 1 - \omega(\rho) - \lfloor \log(\rho) \rfloor]$, and the T-depth is $TD_3 \cdot (\lfloor \log \rho \rfloor + \lfloor \log \frac{\rho}{3} \rfloor + 1)$. Here we do not need any extra ancilla qubit.
- **Step 5.** Here we uncompute Step 2. We just repeat the same Toffolis from Step 2 in reverse order, except for the last one.
  Since we decompose them into Logical-And structures, no additional cost is needed in this step.
- **Step 6.** In step 6, we restore the original binary addends a and b by applying the NOT gate and Toffoli on p and g. In order to store the corresponding b_i in the corresponding qubits, we apply the CNOT gate by taking a_i as the control qubit and b_i as the controlled qubit. Since the Logical-And that used previously introduces the measurement operation, so we then uncompute only the Toffoli gates which are applied to the least significant qubits of each group.
  In this step, the required T-count is $\rho \cdot TC_3$, and the T-depth is TD_3. We do not need any extra ancilla qubits.
- **Step 7.** In this step, we construct the Gidney's RCA [62] to calculate the sum for each group.
  In step 7, the required T-count is $4 \cdot (n - \lceil \frac{n}{r} \rceil)$, the T-depth is r, and the number of extra ancilla qubits is $\alpha - 1 + (r - 2) \cdot \rho$.

An example of an addition operation performed by the higher radix adder is shown in Fig. 3.11. We use seven colors to divide the whole circuit from step 1 to step 7 from left to right. The radix of this adder is set to 3, and the inputs (i.e., addends) are two 15-bit binary numbers denoted by a and b. By using this quantum circuit, we can correctly get the sum of these two numbers. In order to show the overall structure more clearly, we use $\circ$ to represent the controlled-NOT operation of the Logical-And structure in Fig. 3.11.

Interestingly, there are two special cases. When $r \geq n$, our adder is Gidney's RCA. When radix is equal to one, our adder is a simple CLA. In summary, the overall cost of our circuit is shown below.

$$\begin{aligned}\text{T-count} &= (8r+40)\cdot\lceil\tfrac{n}{r}\rceil + 11n - 72 - 7\cdot(n-1)\pmod r - 8r\\ &\quad -21\omega(\lceil\tfrac{n}{r}\rceil-1) - 21\left\lfloor\log(\lceil\tfrac{n}{r}\rceil-1)\right\rfloor)\end{aligned} \tag{3.11}$$

$$\begin{aligned}\text{T-depth} &= 6\cdot(\left\lfloor\log(\lceil\tfrac{n}{r}\rceil-1)\right\rfloor + \left\lfloor\log(\tfrac{1}{3}(\lceil\tfrac{n}{r}\rceil-1))\right\rfloor) + \lfloor\log(r-2)\rfloor\\ &\quad +2r-1\end{aligned} \tag{3.12}$$

$$QC = 3n - 1 - 2r - \omega(\lceil\tfrac{n}{r}\rceil-1) + (2r-1)\cdot\lceil\tfrac{n}{r}\rceil - \left\lfloor\log(\lceil\tfrac{n}{r}\rceil-1)\right\rfloor \tag{3.13}$$

It can be observed that the circuit structure of the quantum higher radix adder varies with radix. In the next section, we will discuss how radix affects the performance of our adder and compare it with other well-known work.

3.1.3 Results and Discussions

3.1.3.1 Experiment 1: The Effect of Radix

Figure 3.12 shows T-count, T-depth, and QC for nine different higher radix adders with radix from 1 to 9, respectively. It is clear that when the radix is fixed, the performance of our adder varies for different input sizes. As the input size increases, the overall cost also increases, which means that the larger the input size, the more complex and expensive any adder tends to be. Since an increase in input size means an increase in the number of operations, this can directly result in an increase in circuit scale. For larger circuit, more expensive cost is often required in terms of T-depth, T-count, and QC.

Interestingly, for a fixed input size, increasing the radix does not always reduce the cost monotonically. In this work, the higher radix adder with r equal to 1 is a CLA, while r equal to 2 represents the higher radix adder without multi-control Toffolis. Compared to higher radix adder with r equal to 1, the cost of it with radix 2 is reduced in T-count, T-depth, and QC, which means that the higher radix layer can effectively optimize the circuit even without introducing multi-control Toffoli. When r is larger than 2, our adder is a hybrid of quantum RCA and CLA. T-count and T-depth decrease first and then increase as the radix increases in the range where r is less than the input size. Meanwhile, QC increases steadily in the fluctuation.

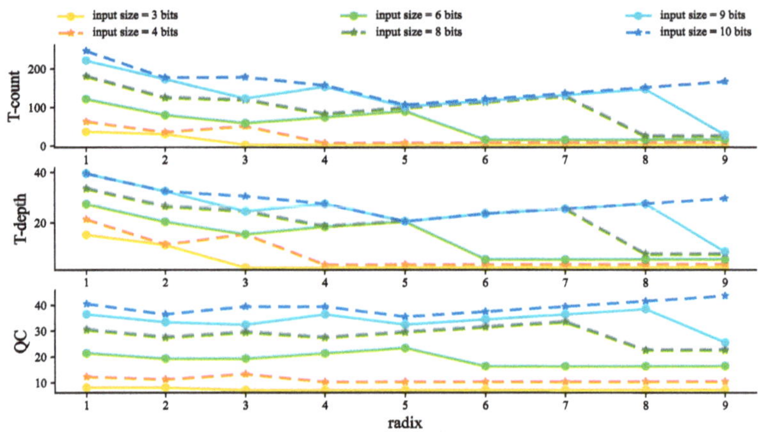

Fig. 3.12 Comparison of the cost required by quantum higher radix adders with different radix and different sizes

When r is equal to input size, QC drops abruptly. This dramatic change is caused by the transformation of our higher radix adder from the hybrid to Gidney's RCA.

It can be seen that the performance of the higher radix adder is significantly influenced by the choice of radix. Therefore, by changing it, we can adapt the proposed adder to the specific requirements of different scenarios as well as minimize the overall cost. In general, when QC is more expensive, a small radix should be set to avoid introducing too much ancillas, while when T-depth or T-count is more costly, setting a larger radix can help minimize T-depth or T-count.

3.1.3.2 Selection of Best Radix

The best radix for T-depth, T-count, and QC is defined as the radix that leads to the lowest cost of our adder in terms of T-depth, T-count, and QC, respectively. As shown in Fig. 3.12, the fluctuations of T-depth, T-count, and QC vary as r increases. Hence, the corresponding best radix may be different for T-depth, T-count, and QC. According to the formulae for cost (given in Eqs. (3.11), (3.12), and (3.13)), the corresponding optimum radix can be determined. See supplementary material for more details.

3.1.3.3 Experiment 2: Comparison with Well-Known Quantum Adders

The performance of our adder is compared to other well-known quantum adders as shown below. We summarize the cost formulas of them in Tables 3.4 and 3.5. Based on them, the relevant data is visualized in Fig. 3.13.

Table 3.4 Performance analysis of different quantum adders. To simplify the representation, we compare the costs of different higher radix adders with r from 3 to $n-1$ and then use the lowest cost to represent the performance of our adder. The formula for $\omega(n)$ is $\omega(n) = n - \sum_{y=1}^{\infty} \left\lfloor \frac{n}{2^y} \right\rfloor$, and the range for r is $2 < r \leq n$

Adder	Year	T-count	T-depth	QC
⋆ VBE RCA [218]	1995	$28n - 14$	$12n - 6$	$3n + 1$
⋄ VBE RCA [218]	1995	$8n + 6$	$3n + 4$	$3n + 1$
⋆ Cuccaro RCA [40]	2004	$14n - 7$	$6n - 3$	$2n + 2$
⋄ Cuccaro RCA [40]	2004	$4n + 3$	$n + 2$	$2n + 2$
⋆ Draper In-place CLA [46]	2004	$70n - 21\omega(n) - 21\omega(n-1) - 21\lfloor \log n \rfloor - 21\lfloor \log(n-1) \rfloor - 49$	$24 + 3 \cdot \lfloor \log n \rfloor + 3 \cdot \lfloor \log(n-1) \rfloor + 3 \cdot \lfloor \log \frac{n}{3} \rfloor + 3 \cdot \left\lfloor \log \frac{n-1}{3} \right\rfloor$	$4n - \omega(n) - \lfloor \log n \rfloor$
⋄ Draper In-place CLA [46]	2004	$50n - 11\omega(n) - 21\omega(n-1) - 11\lfloor \log n \rfloor - 21\lfloor \log(n-1) \rfloor - 39$	$15 + 3 \cdot \lfloor \log n \rfloor + 3 \cdot \lfloor \log(n-1) \rfloor + 3 \cdot \lfloor \log \frac{n}{3} \rfloor + 3 \cdot \left\lfloor \log \frac{n-1}{3} \right\rfloor$	$4n - \omega(n) - \lfloor \log n \rfloor$
⋆ Takahashi Adder [207]	2008	$196n$	$90 \log n$	$2n + \frac{3n}{\log n}$
⋆ Takahashi RCA [208]	2009	$14n - 7$	$6n - 3$	$2n + 1$
⋄ Takahashi RCA [208]	2009	$4n + 3$	$n + 3$	$2n + 1$
⋆ Takahashi combination [208]	2009	$49n$	$54 \log n$	$2n + \frac{3 \cdot n}{\log n}$
⋆ Wang RCA [223]	2016	$7n$	n	$3n + 1$
⋄ Gidney RCA [62]	2018	$4n - 4$	n	$3n - 1$
⋆ Gayathri RCA [180]	2021	$7n$	n	$3n + 1$
⋄ Gayathri Adder [178]	2022	$4n$	$2n$	$3n$
⋆ Our Adder		$56n - 7\frac{n}{r} - 7 \cdot (n-1) \pmod r - 21\omega(\frac{n}{r}) - 21 \log n + 21 \log r - 21$	$12 \log n + 9r - 6 \log r - 6 \log 3r + 6 \log(r-2) + 2$	$4n - \log n + \frac{n}{r} - \omega(\frac{n}{r}) + \log r - 1$
• Our Adder		$46n + 3\frac{n}{r} - 7 \cdot (n-1) \pmod r - 21\omega(\frac{n}{r}) - 21 \log n + 21 \log r - 11$	$12 \log n + 7r - 6 \log r - 6 \log 3r + 6 \log(r-2) + 5$	$4n - \log n + \frac{n}{r} - \omega(\frac{n}{r}) + \log r - 1$
⋄ Our Adder		$19n + 40\frac{n}{r} - 7 \cdot (n-1) \pmod r - 21\omega(\frac{n}{r}) - 21 \log n + 21 \log r - 11$	$12 \log n + 2r - 6 \log r - 6 \log 3r + \log(r-2) + 6$	$5n - \log n - \frac{n}{r} - \omega(\frac{n}{r}) + \log r - 1$

Table 3.5 Performance analysis of quantum out-of-place CLAs. Since Draper's out-of-place adder does not need to be complexly uncomputed like the in-place adder, it is unfair to compare our adder directly to it. Therefore, we construct a simplified version of our higher radix adder to compare with. The relevant formulas are shown below

Adder	Year	T-count	T-depth	QC
⋆ Draper out-of-place CLA [46]	2004	$35n - 21\omega(n) - 21\lfloor \log n \rfloor - 7$	$12 + 3 \cdot \lfloor \log n \rfloor + 3 \cdot \lfloor \log \frac{n}{3} \rfloor$	$4n + 1 - \omega(n) - \lfloor \log n \rfloor$
⋄ Draper out-of-place CLA [46]	2004	$25n - 11\omega(n) - 11\lfloor \log n \rfloor - 7$	$7 + 3 \cdot \lfloor \log n \rfloor + 3 \cdot \lfloor \log \frac{n}{3} \rfloor$	$4n + 1 - \omega(n) - \lfloor \log n \rfloor$
⋆ Our adder		$35n - 7 \cdot (n-1) \pmod r + 7 \cdot n/r - 14\omega(\frac{n}{r}) - 14 \log n + 14 \log r - 14$	$6 \log n + 3r - 3 \log r - 3 \log 3r + 3 \log(r-2) + 5$	$4n - \log n + \frac{n}{r} - \omega(\frac{n}{r}) + \log r - 1$
• Our adder		$32n + 10\frac{n}{r} - 7 \cdot (n-1) \pmod r - 14\omega(\frac{n}{r}) - 14 \log n + 14 \log r - 11$	$6 \log n + 4r - 3 \log r - 3 \log 3r + 3 \log(r-2) + 5$	$4n - \log n + \frac{n}{r} - \omega(\frac{n}{r}) + \log r - 1$
⋄ Our adder		$19n + 26\frac{n}{r} - 7 \cdot (n-1) \pmod r - 14\omega(\frac{n}{r}) - 14 \log n + 14 \log r - 11$	$6 \log n + 2r - 3 \log r - 3 \log 3r + \log(r-2) + 3$	$5n - \log n - \frac{n}{r} - \omega(\frac{n}{r}) + \log r - 1$

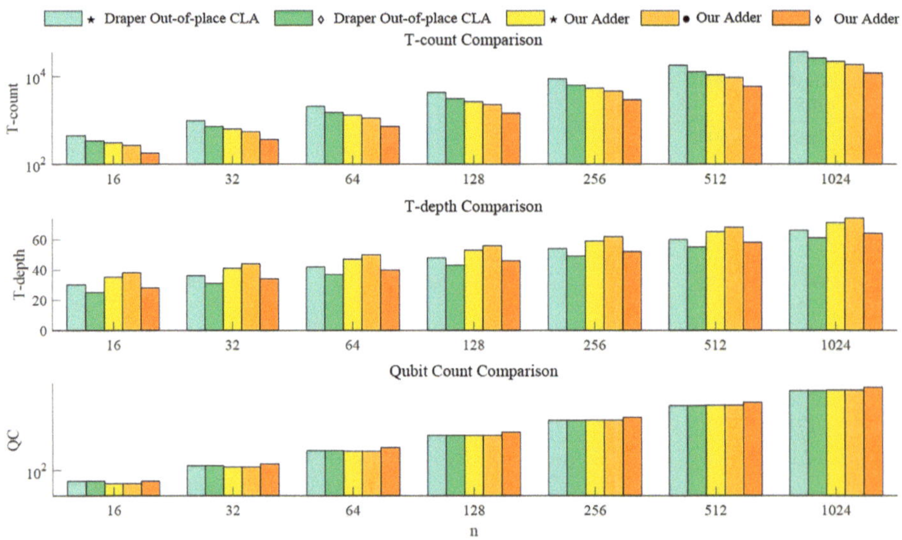

Fig. 3.13 Comparison of the cost required by Draper's out-of-place CLAs and our adders

Firstly, we describe some important experimental details. In order to evaluate our adder more objectively, three Toffoli decomposition methods are used to decompose adders into three different versions. For adders which are denoted by ⋄, all the

Toffoli pairs are decomposed using the Logical-And structure, and then the rest are decomposed using Method 3. For adders which are denoted by $\star$, all the Toffolis are decomposed by Method 3. For adders which are denoted by $\bullet$, only Gidney's RCAs [62] are decomposed using the Logical-And structure, and the rest are decomposed by Method 3. Among them, the adders which are denoted by $\diamond$ and $\star$ have smaller qubits because that Logical-And structure is not used before the sum path, which means some ancilla can be reused in sum path after the uncomputation. For the adders which are denoted by $\bullet$, the overall T-depth and T-count are reduced at the cost of QC. This is due to that it uses Logical-And structures wherever possible.

Then we compare the proposed adder with other well-known works. Compared to quantum RCAs, our adder consistently has a significant advantage in terms of T-depth, despite having more T-count and QC. It is interesting that various new RCA designs are proposed in recent years. In a comparision conducted by Orts et al. [223], the performance of Wang RCA [223] and Gayathri RCA [180] was evaluated. Moreover, Gayathri et al. [178] utilized a variant of Gidney RCA [62] in their quantum circuit, referred to as Gayathri Adder in this work. Overall, Gidney RCA is observed as the lowest cost in T-count, T-depth, and QC among all these new RCA designs. Therefore, it is widely applied in various quantum designs, such as [95]. Specifically, Gidney RCA achieves a T-count of $4n-4$, lower than Wang RCA, Gayathri RCA, and Gayathri Adder. Moreover, Gidney RCA is equal to the T-depth of Wang RCA and Gayathri RCA, while Gayathri Adder exhibits a slightly higher T-depth of $2n$. For QC, Gidney RCA excels with a value of $3n - 1$, while Wang RCA and Gayathri RCA have $3n + 1$ qubits, and Gayathri Adder has a QC of $3n$. To maintain conciseness, the subsequent comparisons will only focus on the Gidney RCA among all the new RCAs which are proposed after 2016. It is evident that even when compared to Gidney RCA, our adder maintains a significant advantage in T-depth. Compared to quantum CLAs, our adders, which are denoted by $\diamond$ and $\bullet$, have similar T-count and T-depth, but significantly smaller T-count. For our adder, which is denoted by $\star$, it significantly reduces the T-depth and further reduces the T-count at the cost of a slight increase in qubit. Since Draper's out-of-place adder [46] does not require complex uncomputation, unlike the in-place variant, we construct a simplified version of our higher radix adder for a fair comparison. According to Fig. 3.13, our adder slightly increases T-depth and QC, but significantly decreases T-count. Moreover, when compared to Takahashi adder [207], which is a special quantum CLA that introduces grouping idea, all the versions of the higher radix adder have similar QC and significantly smaller T-count. For T-depth, our adders, which are denoted by $\diamond$ and $\bullet$, are similar to Takahashi adder, but our adder, which is denoted by $\star$, has a huge reduction. Besides, the higher radix adder is also compared with Takahashi combination adder [208], which also combines RCA and CLA. Although our QC is larger, the T-count and T-depth of our adder have different degrees of reduction. It is obvious that our structure is more general and flexible and further improves the overall efficiency.

In general, the higher radix adder needs more qubits as a cost to significantly reduce the overall T-count and T-depth compared to other adders.

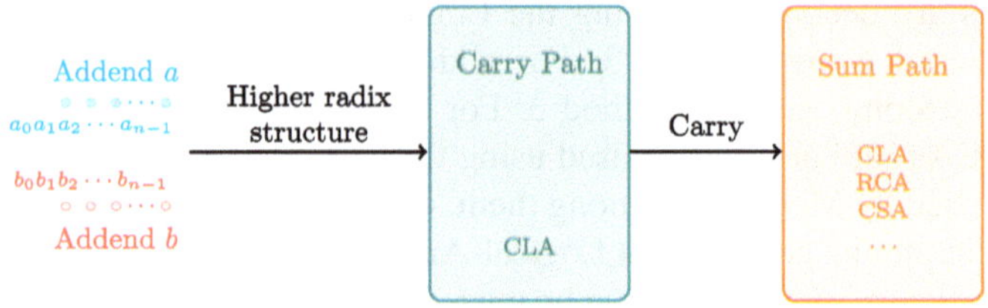

Fig. 3.14 A general framework of higher radix adder

3.1.3.4 Connecting with Existing Quantum Adders

Our work can be seen as a bridge to connect existing quantum adders. Figure 3.14 illustrates the general framework of it, whose key parts are the carry path and the sum path.

For the carry path, quantum CLAs can be used to compute specific carries. For the sum path, any quantum adder can be used to calculate the final result based on those carries. It is interesting to note that the cost contribution of the carry path and sum path in the total circuit can be adjusted by changing radix. More specifically, when the radix is large, the number of groups divided by the higher radix layer is small. Hence, the carry chain is short, which means the sum path is a larger cost contribution of the overall circuit than the carry path. On the contrary, when the radix is small, the carry chain is longer, which means the carry path accounts for a large portion of the total circuit.

According to this general framework, this work can be seen as a specific example based on Draper's CLA and Gidney's RCA. Specifically, our carry path uses the same Brent-Kung tree structure as Draper's CLA, and our sum path is Gidney's RCA. Apart from these two adders, other quantum adders can also be used to construct a higher radix adder.

In order to support one to construct the cheapest quantum adder in different scenarios quickly and easily, we summarize the performance of well-known quantum adders in Fig. 3.15. When QC is more expensive, it is more suitable to use adders with less ancilla such as Takahashi RCA. For T-count, using RCAs such as Gidney's RCA can effectively reduce the overall cost. For reducing T-depth, it is recommended to integrate Draper's In-place CLA or other quantum CLAs within higher radix framework described in Fig. 3.14.

3.1.4 Section Summary

In this study, we thoroughly explained the proposed quantum higher radix adder, covering the background, detailed methodologies, experimental results, and a comprehensive discussion.

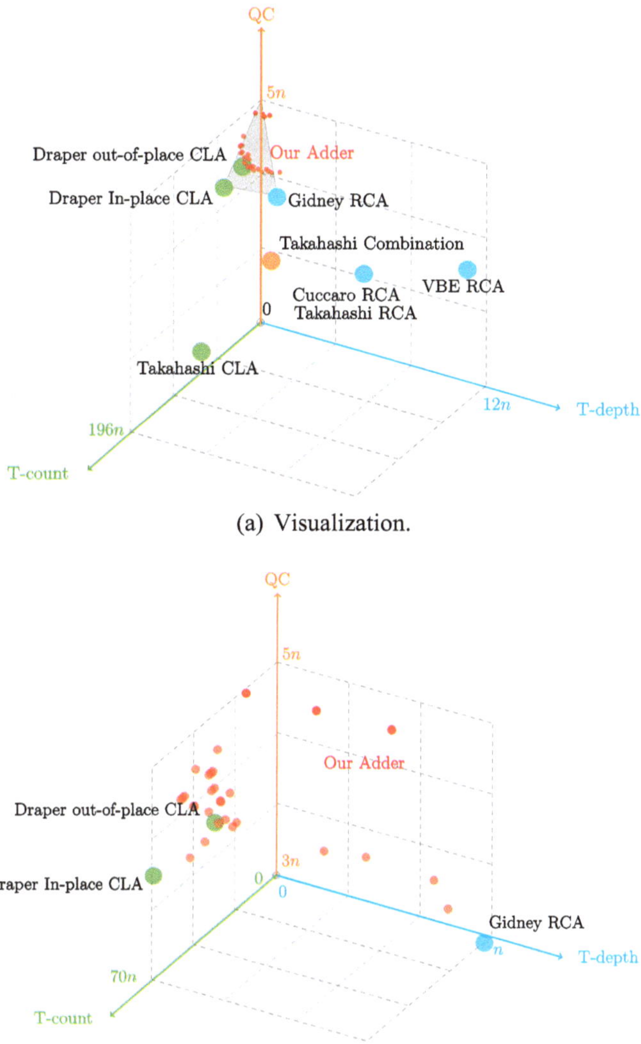

(a) Visualization.

(b) Zoom in on the gray scope.

Fig. 3.15 Visualization of quantum adders

Inspired by classical higher radix adders, we constructed an efficient quantum framework for integer addition by introducing classical higher radix carry look-ahead adders and Manchester carry chain adders in this design. This framework focused on two key components: the carry path and sum path, providing great flexibility for customization according to specific application scenarios.

In our design, we obtained remarkable results. Particularly in terms of T-depth and T-count, our quantum higher radix adder outperformed all the existing quantum carry look-ahead adders, except for Draper's out-of-place CLA. For T-count, the higher radix adder demonstrated a significant reduction compared to Draper's out-of-place CLA, while maintaining comparable qubit count and T-depth.

This in-depth exploration not only enriches the understanding of higher radix adders but also provides valuable insights for future research in quantum area.

3.2 Addition: Quantum Ling Adder

Improving the performance of quantum adder is an important technical challenge with major impact on the implementation of efficient, large-scale quantum computing. Continuing along this research direction, we propose a novel parallel prefix quantum adder based on Ling expansion [226]. We systematically explored classical structures for parallel prefix adders assessing their suitability to be realized in quantum domain. Furthermore, Ling adder enforces Logical OR and large fan-out, which require innovative solutions. We addressed these challenges to realize the quantum Ling adder, which results in a T-depth of only $O(\log \frac{n}{2})$. This represents a substantial improvement over the previous quantum adders based on parallel prefix structure, which require $O(\log n)$ T-depth. We present extensive theoretical and simulation-based studies to establish our claims.

3.2.1 Introduction

In classical CMOS technology, arithmetic circuits have been studied for a long time. In a notable contribution, Ling [128] proposed a different expansion of the parallel prefix calculation, resulting in a reduced depth. However, this structure has not been exploited in quantum adders yet. This chapter proposes a novel quantum adder based on Ling structure, which has the potential to outperform existing structures in T-depth. By thoroughly exploring the unique properties of Ling structure in the quantum world, we develop a quantum Ling adder that significantly reduces resource demands. We encountered several challenges in this research. One significant challenge is the noncopyability of qubits, which posed difficulties in directly translating Ling structure into quantum circuits. Hence, we modified the classical Ling structure, maximizing parallel structures while considering quantum circuit properties. Moreover, our building blocks are refined based on the parallel prefix adder design to address the strict requirements of OR logic in Ling structure.

Our main contribution is the exploration of the effectiveness and potential advantages of Ling structure in the quantum world, compared to existing adder designs. Through a comprehensive evaluation of the proposed design by using T-depth (TD), qubit count (QC), and T-count (TC), we provide several valuable insights into the strengths and limitations of quantum Ling adder, thereby contributing to the continuous progress of quantum computing field.

3.2.2 Integrating Ling into Parallel Prefix

Parallel prefix structure and Ling structure have notable similarities as they both require the pre-calculation of propagation variables and use the Brent-Kung structure for propagation to obtain the final sum. However, Ling structure requires different propagation variables, including pl, gl, and d, as shown in Formula (3.14), Formula (3.15), and Formula (3.16). After performing Ling pre-calculation on pl and gl, the corresponding propagation variables p and g can be obtained.

$$pl_i = a_i + b_i \tag{3.14}$$

$$gl_i = a_i \cdot b_i \tag{3.15}$$

$$d_i = a_i \oplus b_i . \tag{3.16}$$

Moreover, there are differences in the propagation mechanisms between these two structures. In contrast to the parallel prefix structure, Ling structure uses two parallel Brent-Kung structures, as illustrated in Figs. 3.16a and b.

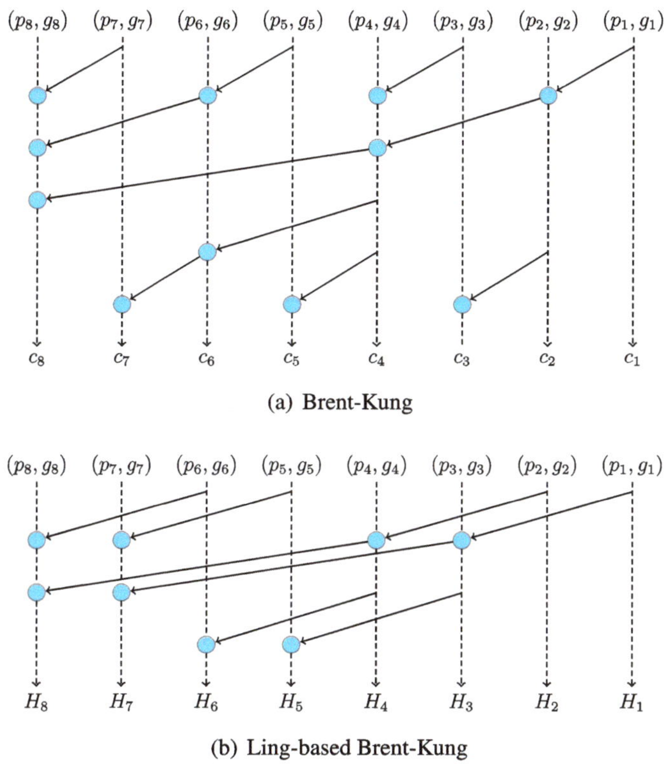

Fig. 3.16 Different Brent-Kung structures

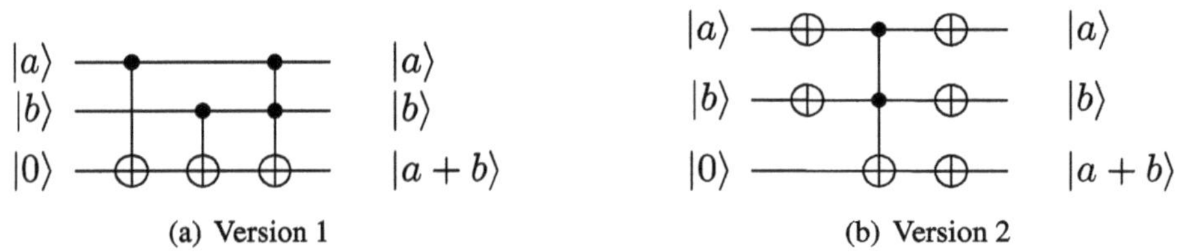

Fig. 3.17 Quantum or gate

In this section, we only focus on the basic building blocks, whereas the subsequent section will cover other implementation details of the quantum Ling structure. Compare with parallel prefix structure, Ling structure has stricter requirements on the OR logic. Specifically, the propagation variables p and g can take four possible values: "00," "01," "10," and "11." Hence, the OR logic here cannot be simplified to XOR logic. As shown in Figs. 3.17a and b, we propose two different quantum OR gates. Based on this work [124], version 2 is used in our design considering the higher cost of two CNOT gates compared to five NOT gates. As mentioned above, apart from the AND and XOR logics, the basic building blocks of our adder also include the quantum OR logic, which is implemented using NOT and Toffoli gates in this work. Most importantly, the Ling structure demonstrates the remarkable potential to outperform the standard parallel prefix structure both in classical and quantum implementation.

3.2.3 *Methodology*

In this section, we describe the proposed quantum Ling adder in detail. To begin with, we present the implementation details of each step, followed by a general overview of the entire design.

3.2.3.1 Step 1

Firstly, we construct few fundamental blocks for the quantum Ling adder, specifically targeting the computation of pl, gl, and d. It is one of the most basic subcircuits of our adder.

Taking a one-bit addition of a and b as an example, the quantum circuit shown in Fig. 3.18 uses two ancilla qubits. Specifically, we apply a Toffoli gate with a and b as control qubits and the first ancilla as the target bit to calculate and store the corresponding gl. To calculate and store the corresponding gl, a Toffoli gate is applied with a and b as control qubits and the first ancilla as the target qubit. Subsequently, employing three CNOT gates with the second ancilla as the target qubit and a, b, and gl as control qubits allows the computation of pl. The resulting

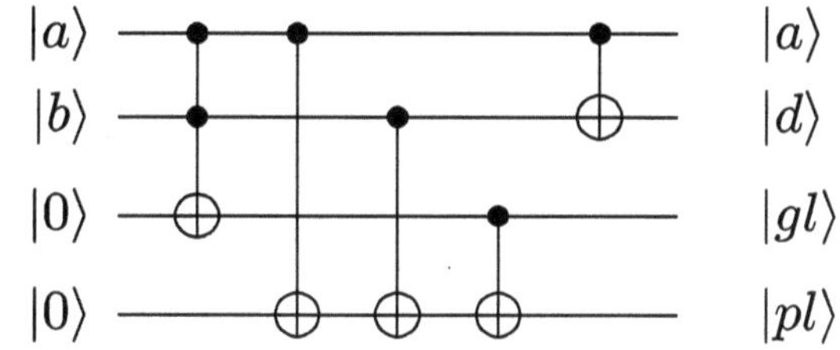

Fig. 3.18 Calculation of pl, gl, and d

value of pl is stored in the second ancilla. This approach eliminates the requirement for quantum OR operation, thereby avoiding the requirement for an extra Toffoli gate. Furthermore, d can be obtained by applying a single CNOT gate with a as the control qubit and b as the target qubit. The final state of d is stored in the qubit which initially stored b.

3.2.3.2 Step 2

In this step, we perform a pre-calculation on pl and gl which are calculated in step 1 to obtain the propagation variables p and g. The corresponding equations are provided in (3.17) and (3.18).

$$g_i = gl_i + gl_{i-1}, \quad g_0 = gl_0 \tag{3.17}$$

$$p_i = pl_i \cdot pl_{i-1}, \quad p_0 = p_1 = 0. \tag{3.18}$$

As shown in Fig. 3.19a, the classical Ling pre-calculation cannot be directly implemented in the quantum world since the inability to copy qubits as classical bits. Hence, the pre-calculation is divided into two parts, illustrated in Figs. 3.19b and c, respectively. This design maximizes the parallel structure, while accommodating the inherent properties of qubits, thereby enhancing the overall efficiency. The pre-calculation circuit in Fig. 3.20 is an example of performing pre-calculation to obtain p_{i-1} and g_i for the i-th addend. It consists of two Toffoli gates and five NOT gates. It is noteworthy that the control and target qubits of these two Toffolis are nonoverlapping, allowing for simultaneous operations within a single time step.

3.2.3.3 Step 3

In the third step, the propagation operations are performed on p and g from the previous step to calculate the intermediate variable H, which is crucial in the summation process.

Circuits for P and G propagations are illustrated in Fig. 3.21. Following Ling-based Brent-Kung structure and using the propagation formulas described in Formula (3.19), we obtain the propagation results which are denoted as P_{i-1} and

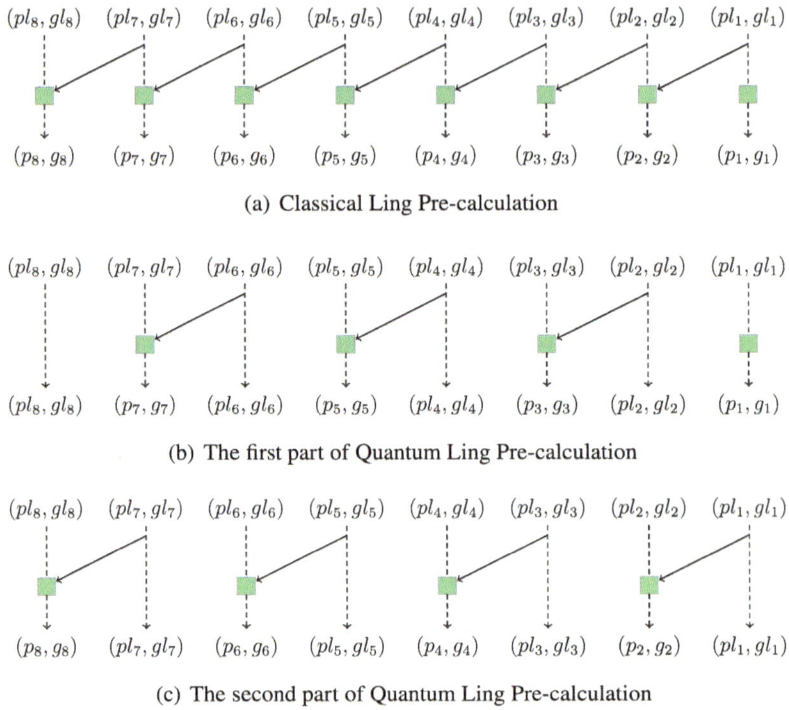

Fig. 3.19 Ling pre-calculation

Fig. 3.20 Pre-calculation of quantum Ling adder

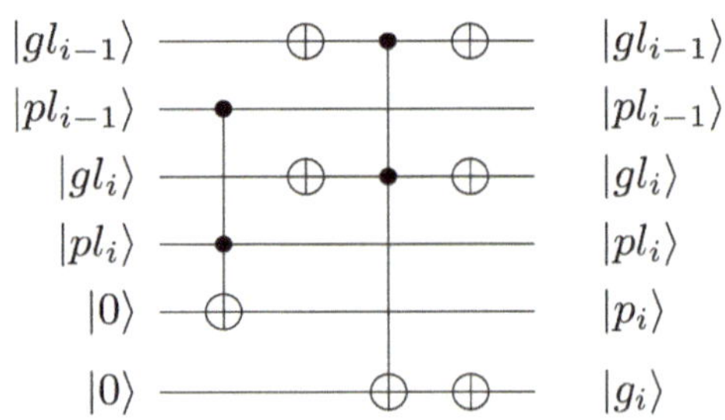

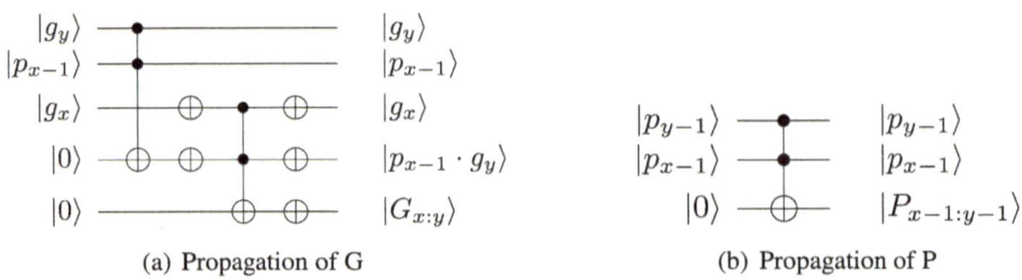

Fig. 3.21 Ling Propagation

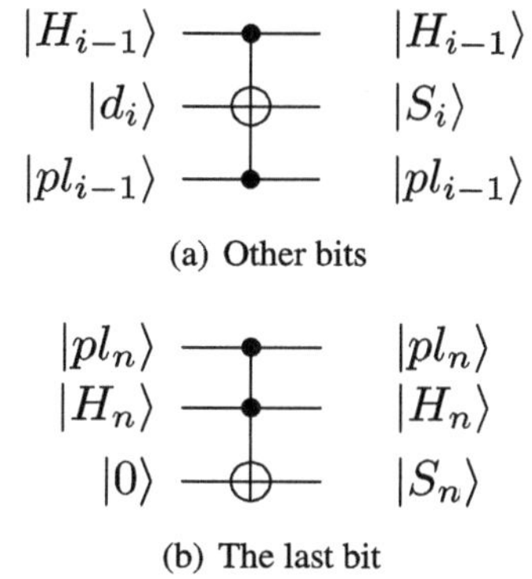

Fig. 3.22 Summation circuit

G_i for the i-th addend, representing the final values of p and g, respectively. Based on P_{i-1} and G_i, we proceed to calculate the intermediate variable H. Especially, H_0 equals to g_0 as demonstrated in Formula (3.20), while the formula for H in other positions is provided by Formula (3.21).

$$g_x, p_{x-1} \circ g_y, p_{y-1} = g_x + p_{x-1} \cdot g_y, p_{x-1} \cdot p_{y-1} \tag{3.19}$$

$$H_0 = g_0 \tag{3.20}$$

$$H_i = G_i + P_{i-1} \cdot G_{i-2}. \tag{3.21}$$

3.2.3.4 Step 4

In this step, we proceed to calculate the final sum, denoted as S. The calculation follows the formulas presented in Formulas (3.22), (3.23), and (3.24).

$$S_0 = d_0 \tag{3.22}$$

$$\begin{aligned} S_i &= \overline{H_{i-1}} \cdot d_i + H_{i-1} \cdot (d_i \oplus pl_{i-1}) \\ &= (H_{i-1} \cdot pl_{i-1}) \oplus d_i \quad , if \quad 0 < i < n \end{aligned} \tag{3.23}$$

$$S_n = pl_{n-1} \cdot H_{n-1}. \tag{3.24}$$

For the first bit of the sum S_0, it is equivalent to d_0. For the intermediate qubits of the sum, a circuit which serves as an illustrative example for comprehending the calculation process is provided in Fig. 3.22a, only using a Toffoli gate. For the last qubit of S, as shown in Fig. 3.22b, it is obtained by using only one Toffoli.

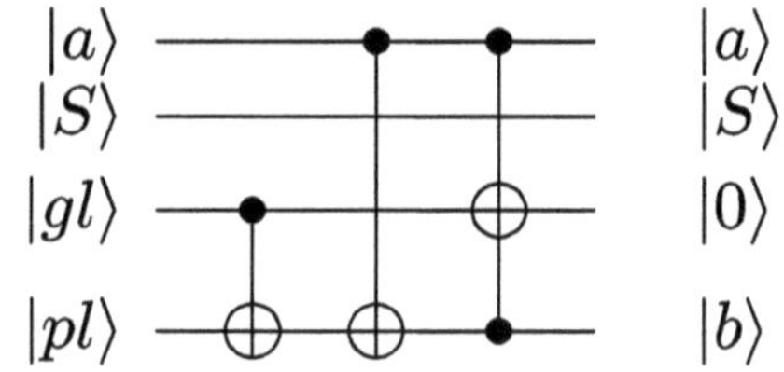

Fig. 3.23 Uncomputation of p, g, and d

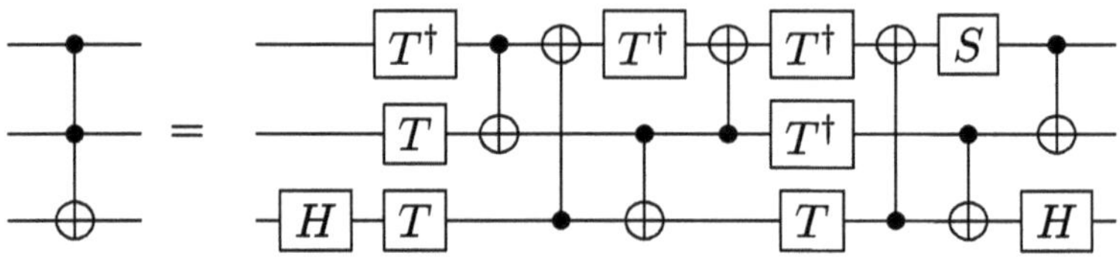

Fig. 3.24 Decomposition method of Toffoli gate [189]

3.2.3.5 Uncomputation

In the fifth and sixth steps, we perform the uncomputation of the quantum circuit. This process is important for the resetting and subsequent reuse of ancilla qubits. We aim to maximize the number of uncomputation operations while preserving the final sum S.

In the fifth step, the circuit is constructed by simply reversing the order of the quantum gated applied in the second step. In the sixth step, as shown in Fig. 3.23, we perform the uncomputation operations for the first step. Particularly, the four qubits in this figure correspond one by one to the four qubits in Fig. 3.18, arranged from top to bottom.

3.2.3.6 Overall Structure of Quantum Ling Adder

Decomposition Method

Overall Structure Our work mainly focuses on the T gate, given its significantly higher cost compared to other quantum gates. Interestingly, Toffoli gate contains T gates. Consequently, a low-cost decomposition method is employed to effectively decompose all the Toffoli gates in our work. For the decomposition, the method proposed by Selinger [189] is used in our work. As shown in Fig. 3.24, the corresponding TD and TC are 3 and 7, respectively.

By combining the six steps introduced previously, a comprehensive and general construction method is proposed for the quantum Ling adder. A comprehensive example of a complete 4-bit quantum Ling adder is depicted in Fig. 3.25. In the previous subsections, the specific quantum circuits for each step have been provided

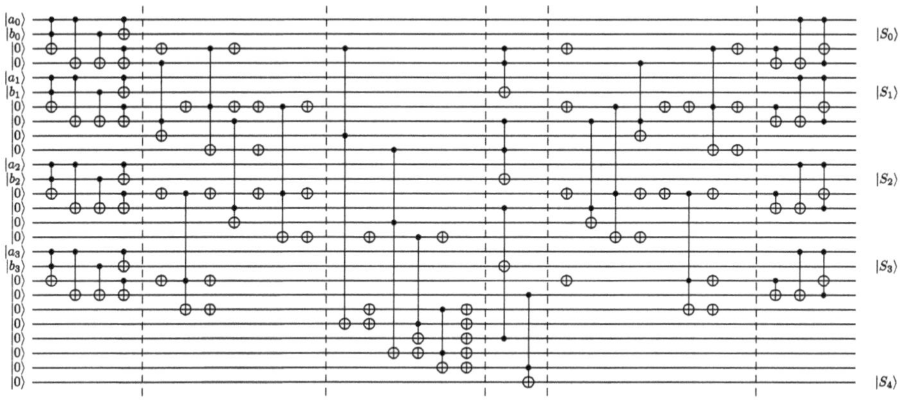

Fig. 3.25 4-bit quantum Ling adder

in detail. Here, a succinct overview of these steps will be presented, accompanied by a comprehensive cost analysis based on TD, TC, and QC.

Step 1 involves the calculation of pl, gl, and d. For n-bit binary addition, the corresponding TD is 3, TC is $7n$, and QC is $4n$. In Step 2, Ling pre-calculation is performed on pl and gl by using a specific method. For n-bit binary addition, the cost analysis indicates a TD of 6, TC of $14n-21$, and a requirement of $2n-3$ extra ancilla. Proceeding to Step 3, Ling propagation is implemented based on the Brent-Kung structure to obtain the intermediate variable H. The cost analysis is as follows: Initially, in the n-bit Brent-Kung structure, a total of $2n - 1 - \omega(n) - \lfloor \log(n) \rfloor$ propagation operations are performed with $\lfloor \log(n) \rfloor + \lfloor \log \frac{n}{3} \rfloor + 1$ logical levels, where $\omega(n)$ represents $n - \sum_{y=1}^{\infty} \lfloor \frac{n}{2^y} \rfloor$. In Ling-based Brent-Kung structure, the initial structure is extended by incorporating two parallel Brent-Kung structures. $(2n-2-2\omega(\frac{n}{2})-2\lfloor \log(\frac{n}{2}) \rfloor)$ propagation operations are executed, with $\lfloor \log \frac{n}{2} \rfloor + \lfloor \log \frac{n}{6} \rfloor + 1$ logical levels. The proposed step 3 is designed based on Ling-based Brent-Kung tree structure. The corresponding TD is obtained by multiplying the logical levels of Ling-based Brent-Kung structure by 3 TD for each Toffoli, resulting in $6+6\lfloor \log \frac{n}{2} \rfloor + 6\lfloor \log \frac{n}{6} \rfloor$. During propagation, the logical levels of P propagation are two less than those of G propagation. To simplify the formulas, we assume an additional layer of P propagation. Hence, upper bounds for TC and QC can be obtained. Specifically, the upper bound for TC is $42n-56-42\omega(\frac{n}{2})-42\lfloor \log \frac{n}{2} \rfloor$, and the number of extra ancilla is $6n-8-6\omega(\frac{n}{2})-6\lfloor \log \frac{n}{2} \rfloor$. In Step 4, the intermediate variable H is utilized to compute the final sum. For n-bit binary addition, the corresponding TD is 3, TC is $7n$, and one extra ancilla is required. Step 5 involves the uncomputation of Step 2. Similar to Step 2, the corresponding TD is 6, TC is $14n - 21$, but no additional ancilla is needed. In the last step, the uncomputation of Step 1 is performed to reset certain ancilla to the initial state $|0\rangle$ with a TD of 3, TC of $7n$, and no extra ancilla requirement.

$$
\begin{aligned}
\text{TD} = &\overbrace{3}^{\text{Step 1}} + \overbrace{6}^{\text{Step 2}} + \overbrace{6 + 6\lfloor\log\frac{n}{2}\rfloor + 6\lfloor\log\frac{n}{6}\rfloor}^{\text{Step 3}} \\
&+ \overbrace{3}^{\text{Step 4}} + \overbrace{6}^{\text{Step 5}} + \overbrace{3}^{\text{Step 6}} \\
=&27 + 6\lfloor\log\frac{n}{2}\rfloor + 6\lfloor\log\frac{n}{6}\rfloor
\end{aligned}
\tag{3.25}
$$

$$
\begin{aligned}
\text{TC} = &\overbrace{7n}^{\text{Step 1}} + \overbrace{14n - 21}^{\text{Step 2}} + \overbrace{42n - 56 - 42\omega(\frac{n}{2})}^{\text{Step 3}} \\
&\overbrace{-42\lfloor\log\frac{n}{2}\rfloor}^{\text{Step 3}} + \overbrace{7n}^{\text{Step 4}} + \overbrace{14n - 21}^{\text{Step 5}} + \overbrace{7n}^{\text{Step 6}} \\
=&91n - 42\omega(\frac{n}{2}) - 42\lfloor\log\frac{n}{2}\rfloor - 98
\end{aligned}
\tag{3.26}
$$

$$
\begin{aligned}
\text{QC} = &\overbrace{4n}^{\text{Step 1}} + \overbrace{2n - 3}^{\text{Step 2}} + \overbrace{6n - 8 - 6\omega(\frac{n}{2}) - 6\lfloor\log\frac{n}{2}\rfloor}^{\text{Step 3}} \\
&+ \overbrace{1}^{\text{Step 4}} + \overbrace{0}^{\text{Step 5}} + \overbrace{0}^{\text{Step 6}} \\
=&12n - 6\omega(\frac{n}{2}) - 6\lfloor\log\frac{n}{2}\rfloor - 10.
\end{aligned}
\tag{3.27}
$$

As mentioned above, the total cost of these six steps can be obtained by summing the individual TDs, TCs, and QCs, as represented by (3.25), (3.26), and (3.27), respectively. For n-bit adder, the total TD, TC, and QC for the entire circuit are $27 + 6\lfloor\log\frac{n}{2}\rfloor + 6\lfloor\log\frac{n}{6}\rfloor$, $91n - 42\omega(\frac{n}{2}) - 42\lfloor\log\frac{n}{2}\rfloor - 98$, and $12n - 6\omega(\frac{n}{2}) - 6\lfloor\log\frac{n}{2}\rfloor - 10$, respectively.

It can be observed that the quantum Ling adder has been extended and improved based on the quantum parallel prefix adder. In the next section, we will discuss the performance of our adder and compare it with other prominent works.

3.2.4 Results and Discussions

In this section, we describe the performance of the proposed quantum Ling adder and provide comprehensive discussions.

For the experimental implementation, the quantum Ling adder is implemented using Qiskit 0.23.3 and Qasm Simulator. The relevant code is available as a

Table 3.6 Performance analysis of different quantum in-place adders. The formula for $\omega(n)$ is $\omega(n) = n - \sum_{y=1}^{\infty} \lfloor \frac{n}{2^y} \rfloor$, where n denotes the bit width of the addend

Adder	Year	TC	TD	QC
Draper in-place CLA [46]	2004	$70n - 21\omega(n) - 21\omega(n-1) - 21\lfloor\log n\rfloor - 21\lfloor\log(n-1)\rfloor - 49$	$24 + 3\lfloor\log n\rfloor + 3\lfloor\log(n-1)\rfloor + 3\lfloor\log\frac{n}{3}\rfloor + 3\lfloor\log\frac{n-1}{3}\rfloor$	$4n - \omega(n) - \lfloor\log n\rfloor$
Takahashi adder [207]	2008	$196n$	$90\log n$	$2n + \frac{3n}{\log n}$
Takahashi combination [208]	2009	$49n$	$54\log n$	$2n + \frac{3\cdot n}{\log n}$
Higher radix adder [224]	2023	$56n - 7\frac{n}{r} - 7\cdot(n-1) \pmod r - 21\omega(\frac{n}{r}) - 21\log n + 21\log r - 21$	$12\log n + 9r - 6\log r - 6\log 3r + 6\log(r-2) + 2$	$4n - \log n + \frac{n}{r} - \omega(\frac{n}{r}) + \log r - 1$
Our adder	–	$91n - 42\omega(\frac{n}{2}) - 42\lfloor\log\frac{n}{2}\rfloor - 98$	$27 + 6\lfloor\log\frac{n}{2}\rfloor + 6\lfloor\log\frac{n}{6}\rfloor$	$12n - 6\omega(\frac{n}{2}) - 6\lfloor\log\frac{n}{2}\rfloor - 10$

public repository.[3] Our design exhibits remarkable scalability of the quantum Ling structure, which enables automatic construction of adders with large bit sizes, while respecting the technology constraints such as limited fan-out.

In Table 3.6, we present a comprehensive comparison of different quantum adders. The corresponding charts of these comparisons are presented in Figs. 3.26a, b, and c. It is worth highlighting that the bit length of the addend, denoted as n in this work, significantly influences the adder's performance. As n increases, the TC, TD, and QC of all the adders tend to increase accordingly. Draper in-place CLA based on parallel prefix structure initially achieves TD of $O(\log n)$, accompanied by TC complexity of $O(70n)$ and QC of $O(4n)$. Based on Draper in-place CLA, Takahashi adder introduces specific modifications. Thereby, the corresponding QC decreases to $O(2n)$, while increasing TD and TC. Further enhancing the performance, Takahashi combination adder combines parallel prefix structure with ripple carry structure. This optimization allows it to maintain low QC of $O(2n)$, while significantly reducing TC and TD by nearly half compared to Takahashi adder. Recently, Wang [224] et al. proposed a dynamic approach to fine-tune the quantum CLA, resulting in improved TD optimization. For the clarity of this work, its performance is not shown in the figures, considering that the parameter r which denoted radix can take various values. In contrast, our proposed adder offers a significant advantage over the other designs. It stands out with the lowest TD of $O(\log \frac{n}{2})$, while maintaining comparable QC and TC. Thus, especially when there is a strict requirement for low TD, our adder will become the obvious choice.

[3] https://github.com/Siyi-06/Quantum_ling_adder

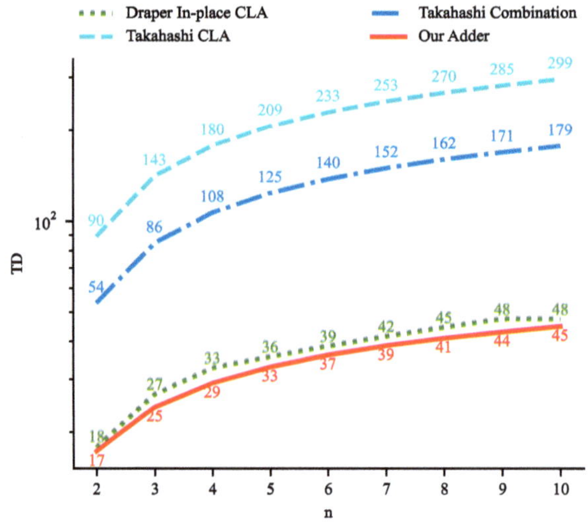

(a) TD Comparison

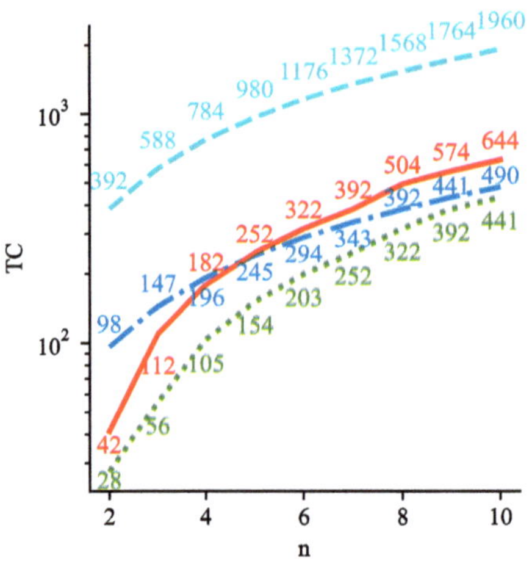

(b) TC Comparison

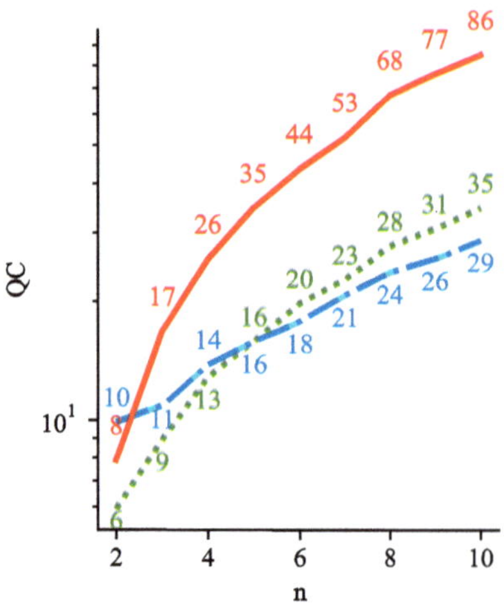

(c) QC Comparison

Fig. 3.26 Comparative cost analysis of quantum Ling adder and other prominent quantum in-place adders

3.2.5 Section Summary

In conclusion, we propose a novel quantum adder based on Ling structure, which reduces depth compared to all the prior quantum adders based on parallel prefix structure. The presented quantum Ling adder achieves a T-depth of $O(\log \frac{n}{2})$. This represents a significant improvement compared to parallel prefix-based adders, including Draper carry look-ahead adder and Takahashi carry look-ahead adders, which has a T-depth of $O(\log n)$.

3.3 Addition: Quantum Optimal Toffoli-Depth Adder

To date, the logarithmic-depth quantum adders include a constant coefficient $k \geq 2$, while achieving the Toffoli-Depth of $k \log n + O(1)$. In this chapter, 160 alternative compositions of the carry-propagation structure are comprehensively explored to determine the optimal depth structure for a quantum adder. By extensively studying these structures, it is shown that an exact Toffoli-Depth of $\log n + O(1)$ is achievable. This presents a reduction of Toffoli-Depth by almost 50% compared to the best-known quantum adder circuits presented to date. We demonstrate a further possible design by incorporating a different expansion of propagate and generate forms, as well as an extension of the modular framework. This work elaborates on these designs, supported by detailed theoretical analyses and simulation-based studies, firmly substantiating our claims of optimality within all possible configurations outlined in this work. The results also mirror similar improvements, recently reported in classical adder circuit complexity.

3.3.1 Introduction

While adder circuits have been frequently explored in classical computing, they remain relatively less chartered in the quantum computing paradigm. Arithmetic circuits have been extensively explored in classical computing. Interestingly, the Sklansky Tree [195] is widely applied for its minimal depth among all the other parallel prefix structures. However, implementing this structure in the quantum world presents significant challenges, primarily due to qubit noncopyability.

This work proposes a novel quantum adder based on the Sklansky Tree, which is verified as the optimal depth structure among all the quantum adders. Throughout this research, we faced a series of challenges. Specifically, the inability to copy qubits presented a barrier to the direct application of the prefix tree adders in the quantum world. To address this issue, we developed a quantum repeat gate that helps to efficiently incorporate the different prefix tree structures such as Sklansky [195] into the quantum world under specific constraints. The proposed repeat gate not only contributes to the establishment of an optimal depth quantum adder but also effectively addresses the problems arising from qubit noncopyability in other specific quantum circuits.

Our primary contribution is the development of the optimal depth quantum adder that achieves a remarkable Toffoli-Depth of $\log n + O(1)$ for n bit-sized additions. Obviously, it marks an important improvement over all the previous quantum carry look-ahead adders based on the Brent-Kung tree, which required a minimum of $2 \log n + O(1)$ Toffoli-Depth. By conducting a thorough assessment involving Toffoli-Depth, qubit count, and Toffoli-Count, this work offers significant insights into the strengths and constraints of the quantum optimal depth adder, enhancing the continuous development of quantum computing, which parallels the development of classical adder analysis [92]. In brief, the main contributions include:

- Thoroughly investigate the prefix tree and explore computation forms using different propagation and generation techniques.
- Propose innovative designs for optimal Toffoli-Depth adders, optimal Toffoli-Depth Ling adder, and optimal Toffoli-Depth modular adder, achieving the peak performance in quantum addition circuits.

3.3.2 Design Choices for Depth Reduction

In this section, we elaborate on the choices to be made with the primary objective of reducing Toffoli-Depth.

As shown in Fig. 3.27, in this chapter, we explored a total of 160 design options. In Sects. 3.3.2.1 and 3.3.2.2, we provided detailed descriptions of the prefix tree and the propagation and generation computation methods, respectively. Regarding addition strategies, there are two options: in-place addition and out-of-place addition, but this work primarily focuses on the quantum out-of-place adder with lower Toffoli-Depth. In Sect. 3.3.3, we discuss the uncomputation strategy, presenting a framework for an adder with uncomputation. Specifically, this framework is outlined by removing all uncomputation steps, such as step 3, to obtain a version without uncomputation. In Sect. 3.3.4.1, two different Toffoli strategies are proposed. Specifically, Strategy 1 involves only Toffoli gates, while Strategy 2 combines Toffoli gates with logical-AND for further optimization of our

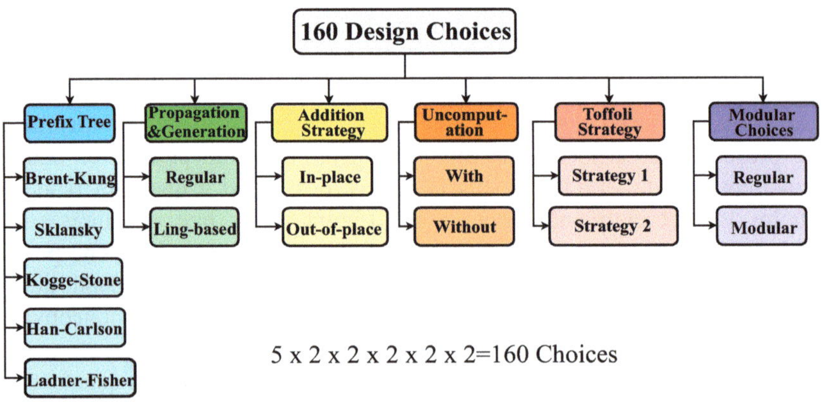

Fig. 3.27 Design choices

Table 3.7 Comparison of dominant prefix trees for n-bit addition

Prefix tree	Logical depth	Operation nodes
Brent-Kung	$2\log n - 2$	$2n - \log n - 2$
Sklansky	$\log n$	$(n/2) * \log n$
Kogge-Stone	$\log n$	$n \log(n/2) + 1$
Han-Carlson	$\log n + 1$	$(n/2) * \log n$
Ladner-Fisher	$\log n + 1$	$\frac{3}{4}n - 1 + (n/4) * \log n$

circuits. For modular choices, we include both regular adders and modular adders, with specific details discussed in Sect. 3.3.4.4.

3.3.2.1 Choice of Prefix Tree

As shown in Table 3.7, the classical Sklansky tree and Kogge-Stone tree stand out as the most efficient parallel prefix structures in terms of logical depth compared to other similar structures, such as the Brent-Kung tree and the Han-Carlson tree.

Among state-of-the-art logarithmic-depth quantum adders, the common approach is to use the Brent-Kung tree [46, 224, 226]. However, it is well-known in the classical adder that the Kogge-Stone and Sklansky prefix trees can achieve a significantly lower logical depth, i.e., $\log n$, reducing depth of the Brent-Kung tree by half. Given this substantial difference in logical depth, it is surprising that these prefix trees were not studied earlier. One possible reason for the same is the prefix tree incurs an overhead due to copy operation, which is inherently prevented in quantum domain due to no-cloning theorem. However, this limitation can be bypassed through introduction of ancilla qubits with initialization states as $|1\rangle$ or $|0\rangle$ and proper application of CNOT gate as shown in Fig. 3.28.

Among the five prominent quantum prefix tree structures, Quantum Sklansky (Fig. 3.28b) stands out as particularly advantageous. While it exhibits the lowest logical depth, Quantum Sklansky also features significantly fewer CNOT operations compared to Quantum Koggle Stone, which has an equivalent logical depth.

3.3.2.2 Choice of Propagate and Generate Computation

The efficiency of quantum carry look-ahead adder depends on the choice of propagate and generate strategy. Here, two common approaches are described as follows:

- **Conventional Propagation and Generation.** The conventional method employs standard logical operations for carry propagation and generation as shown in formulas (3.2), (3.3), (3.4), (3.5), (3.6), and (3.7). It is widely used for its simplicity, providing a clear and straightforward construction of quantum carry look-ahead adders.

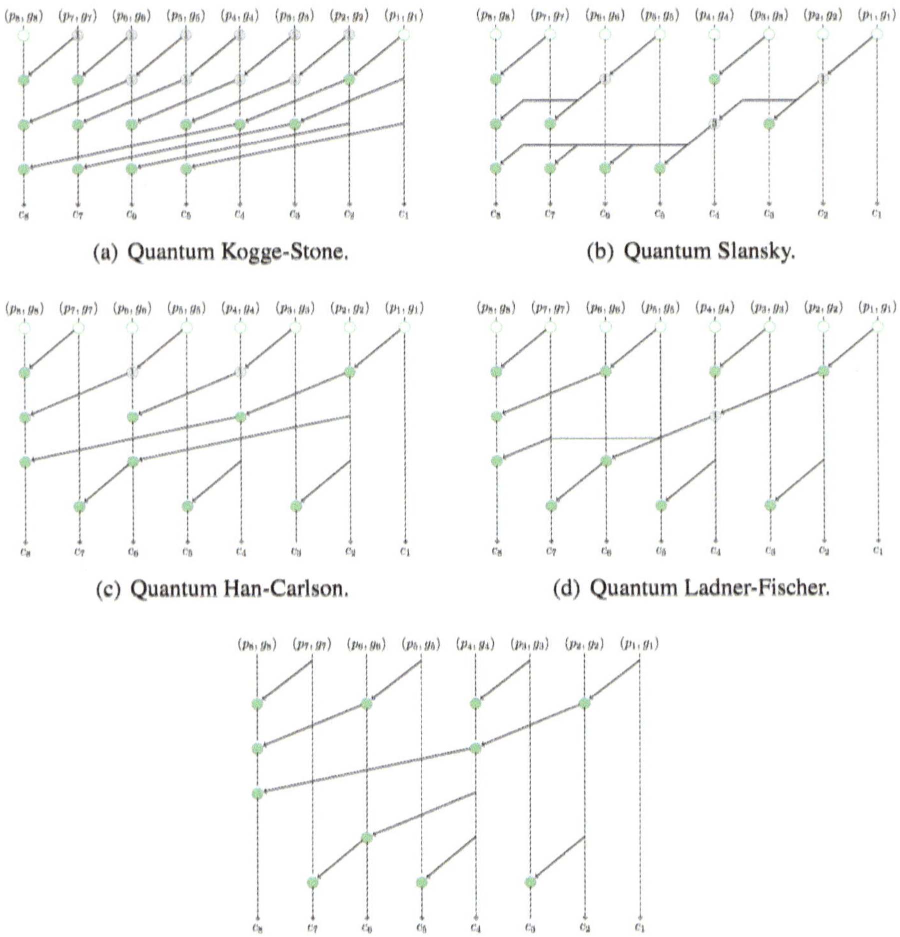

(a) Quantum Kogge-Stone.

(b) Quantum Slansky.

(c) Quantum Han-Carlson.

(d) Quantum Ladner-Fischer.

(e) Quantum Brent-Kung.

Fig. 3.28 Quantum prefix tree. Gray nodes in the prefix trees represent CNOT operations, with numbers in circles indicating the times of CNOT operations

- **Ling-Based Propagation and Generation.** This method is represented by the quantum Ling adder [226], which incorporates the Ling basis [128] into the dominant quantum carry look-ahead adders (CLA) which mainly based on Brent-Kung structure. The Ling-based propagation and generation structure introduces a more complex pre-calculation part, thereby incorporating an additional parallel calculation tree in the calculation process, resulting in a reduction in depth cost. Compared to the traditional approach, the complexity of Toffoli-Depth can be further reduced to $O(\log \frac{n}{2})$ from $O(\log n)$ by integrating the Ling expansion and utilizing a slightly different pre-calculation structure.

Fig. 3.29 Quantum circuit for p_i and g_i calculation

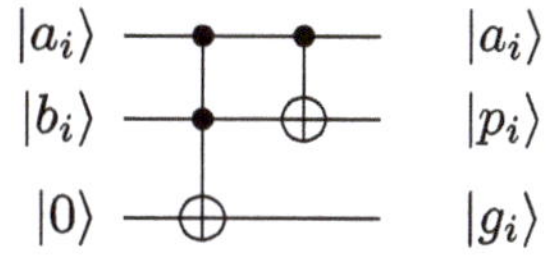

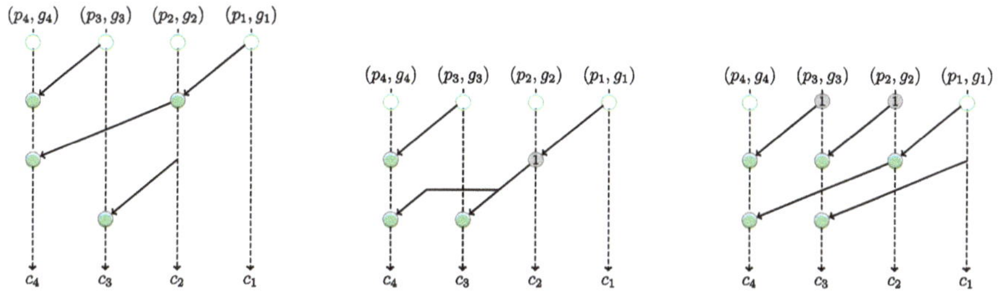

(a) Brent-Kung Prefix Tree. (b) Slansky Prefix Tree. (c) Kogge-Stone Prefix Tree.

Fig. 3.30 4-bit quantum prefix tree. Interestingly, the 4-bit Brent-Kung prefix tree is equivalent to the Han-Carlson and Ladner-Fischer structures

Fig. 3.31 Step 2

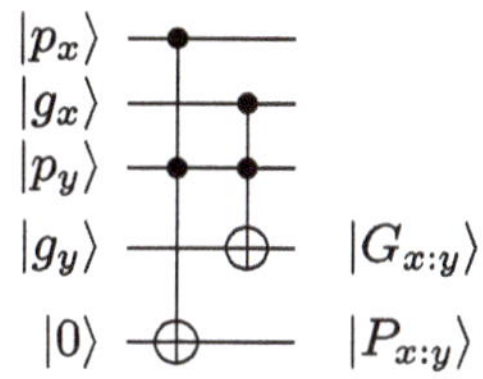

3.3.3 Methodology

In this section, a detailed description of our quantum prefix tree adders is provided (as shown in Fig. 3.29).

As shown in Fig. 3.34, we construct the quantum prefix tree adder by combining the steps 1 to 4:

- **Step 1.** The initial propagation p_i and generation g_i are calculated using a single ancilla qubit and one Toffoli gate (as shown in Fig. 3.30).
  Consequently, Toffoli-Depth is 1, Toffoli-Count is n, and qubit count is $(3n + 1)$.

$$p_i = a_i \oplus b_i \tag{3.28}$$

$$g_i = a_i \cdot b_i . \tag{3.29}$$

- **Step 2.** We implement both P propagation and G propagation (Fig. 3.31) based on the corresponding quantum prefix tree structures as shown in Fig. 3.30.

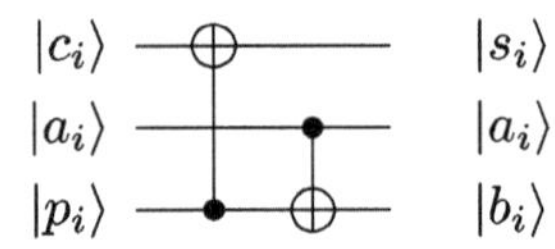

Fig. 3.32 Quantum circuit for computing s from p and c

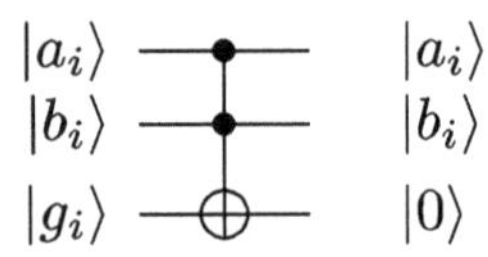

Fig. 3.33 Quantum circuit for uncomputation (only for copied g_i)

The cost of this step depends on the prefix tree structure employed. Specifically, for the quantum Kogge-Stone adder, this step involves a Toffoli-Depth of $(1 + \log n)$, a Toffoli-Count of $(n \log n - \frac{3}{2}n + n \log \frac{n}{2} + 2)$, and an extra qubit count of $(3n \log n - \frac{7}{2}n + 5)$.

- **Step 3.** In this step, P and the copied G are uncomputed based on different prefix tree structures. For example, considering the Kogge-Stone tree, this process incurs a Toffoli-Depth of $(\log n - 1)$. The corresponding Toffoli-Count is $(2n \log n - \frac{7}{2}n + 5)$, and no extra qubit (as shown in Fig. 3.32).
- **Step 4.** This step involves sum calculation and p_i uncomputation without the employment of Toffoli gates.
  Besides, here is no extra qubit required.

$$s_i = p_i \oplus c_i \tag{3.30}$$

$$s_{n+1} = c_{n+1}. \tag{3.31}$$

- **Additional Step.** This step is exclusively designed for the quantum Kogge-Stone structure due to its intricate CNOT operations, necessitating an extra step to ensure the complete uncomputation of all ancillary qubits. As illustrated in Fig. 3.33, this step involves uncomputing all the copied g_i using a Toffoli-Depth of 1 and a Toffoli-Count of $(n - 2)$. Similar to Step 4, this step does not require any ancilla.

The overall structures are depicted in Fig. 3.34, showcasing three distinct examples of 4-bit addition quantum circuits. Interestingly, the equivalence of quantum Brent-Kung, Han-Carlson, and Ladner-Fischer structures is evident within this proposed framework. The comprehensive cost analysis is provided in the subsequent section.

3.3.4 Results and Discussions

In the previous section, we proposed the general framework and specific quantum implementations of several quantum prefix tree adders. In this section, we will

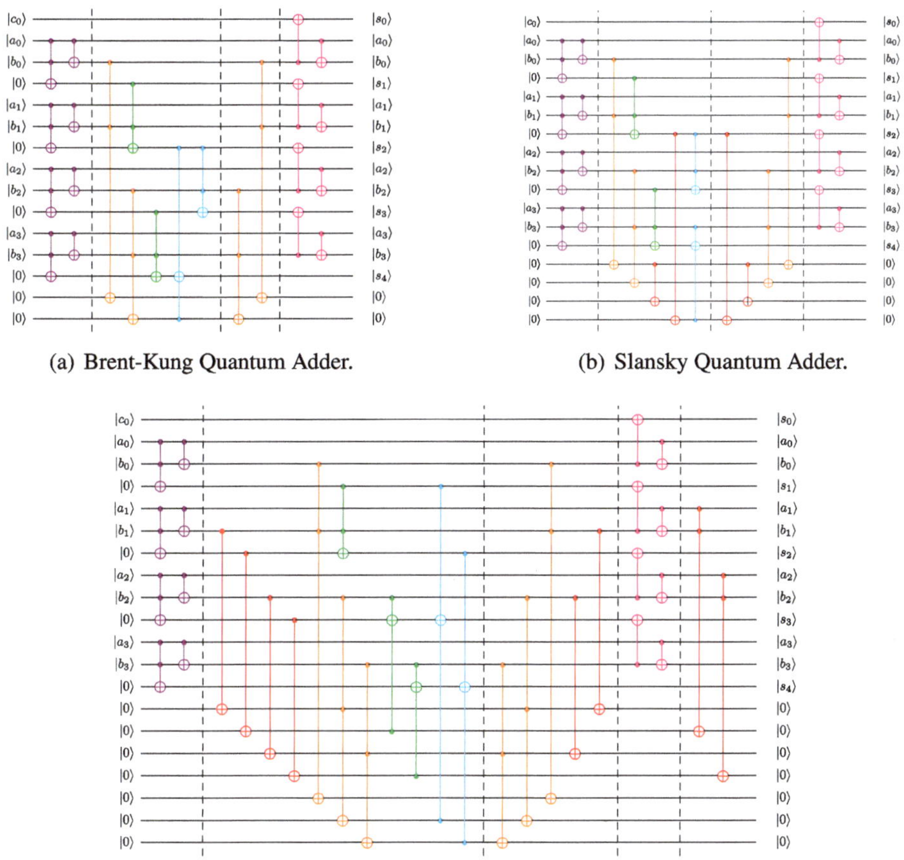

(a) Brent-Kung Quantum Adder.

(b) Slansky Quantum Adder.

(c) Kogge-Stone Quantum Adder.

Fig. 3.34 4-bit quantum adder examples. The dashed lines divide the entire circuit into five parts according to steps 1 to 4. In the first step, purple represents the calculation for initial propagation and generation. In the second step, red represents the copy operation, orange represents the first layer of propagation, and green and blue represent the first and the rest layers of generation, respectively. In the third step, orange represents propagation uncomputation, and red represents the uncomputation of the initial propagation copy p_i. Next, pink represents the operations of the fourth step. In the final step of Kogge-Stone Adder, we perform uncomputation on all the initially copied g_i, represented in red

provide all the details of the results and discussions. Our investigation focuses on the quantum out-of-place adder due to its direct reflection of the prefix tree's influence and its lower Toffoli-Depth compared to the quantum in-place adder. Conversely, while the in-place adder follows a similar workflow, it typically involves more intricate supplementary steps, resulting in a deeper Toffoli-Depth.

Table 3.8 Performance analysis of different quantum adders utilizing Strategy 1

Adder	Toffoli-Count	Toffoli-Depth	Qubit count
Brent-Kung	$5n - 3\omega(n) - 3\lfloor \log n \rfloor - 1$	$4 + \lfloor \log n \rfloor + \lfloor \log \frac{n}{3} \rfloor$	$4n + 1 - \omega(n) - \lfloor \log n \rfloor$
Sklansky	$\frac{3}{2}n \log n + 2\lceil \log n \rceil - n$	$2 \log n + 1$	$n + n \log n + \lceil \log n \rceil + 2$
Kogge-Stone	$3n \log n + n \log \frac{n}{2} - 3n + 5$	$2 \log n + 2$	$3n \log n - \frac{n}{2} + 6$
Han-Carlson	$n + \frac{3}{2}n \log n - 2\lfloor \frac{n}{2} \rfloor$	$2 \log n + 3$	$\frac{3}{2}n + n \log n - \lfloor \frac{n}{2} \rfloor + 3$
Ladner-Fisher	$\frac{13n}{4} + \frac{3n \log n}{4} - 2\lfloor \frac{n}{2} \rfloor - 3$	$2 \log n + 3$	$3n + \frac{n \log n}{2} - \lfloor \frac{n}{2} \rfloor + 1$

3.3.4.1 Identifying the Optimal Depth Adder Among Quantum Prefix Tree Adders

First of all, we conducted a comprehensive analysis of the costs associated with all proposed quantum prefix tree adders to determine the optimal depth structure within the quantum realm. In this process, we introduced two distinct strategies.

Strategy 1
Under Strategy 1, we exclusively employ the Clifford+Toffoli gate set to describe and evaluate the entire quantum addition circuit. Remarkably, the quantum Brent-Kung adder here is equivalent to the existing Draper out-of-place adder [46].

By applying strategy 1, the corresponding costs are detailed in Table 3.8. Unfortunately, it is obvious that compared with the quantum Brent-Kung prefix tree adder, the alternative prefix tree structures we proposed do not demonstrate significant advantages, especially concerning the Toffoli-Depth (as shown in Fig. 3.35).

Strategy 2
In 2018, Gidney proposed a more cost-effective logical AND structure for implementing a pair of Toffoli gates [62]. By utilizing this design, Strategy 2 aims to further optimize quantum adders based on Strategy 1. Specifically, for the proposed quantum prefix tree adder framework, only the P propagation involves Toffoli pairs. Hence, we utilize the logical AND structure for all Toffoli gates in both the P computation and P uncomputation stages. For other unpaired Toffoli gates, we still use Clifford+Toffoli gate set to describe them.

Under Strategy 2, the respective costs can be found in Table 3.9. It is evident that all proposed designs significantly outperform the original quantum Brent-Kung adder in terms of Toffoli-Depth. Particularly, the quantum Sklansky adder exhibits the quantum optimal Toffoli-Depth, which is only $log(n) + 1$.

After comparing all the proposed designs, we confirm that the quantum Sklansky + Strategy 2 is the optimal depth design choice in this work. This is attributed to its minimal Toffoli-Depth and fewer CNOT operations involved, especially considering that Sklansky only requires P-CNOT operations during CNOT operations, without the need for G-CNOT operations.

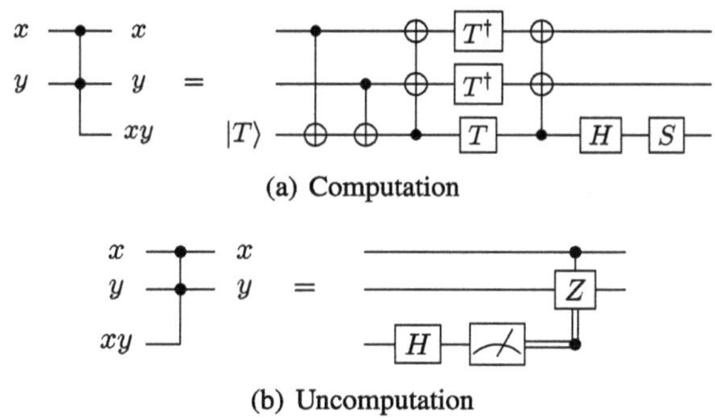

Fig. 3.35 Gidney's Logical-And structure

Table 3.9 Performance analysis of different quantum adders utilizing Strategy 2

Adder	Extra T count	Toffoli count	Extra T-depth	Toffoli depth	Qubit count
Brent-Kung	$8n - 4\log n - 8 - 4\lfloor \frac{n}{2} \rfloor$	$2n - \log n - 2$	2	$2\log n - 1$	$4n + 1 - \omega(n) - \lfloor \log n \rfloor$
Sklansky	$2n\log n - 4n + 4\lceil \log n \rceil$	$\frac{n \log n}{2}$	2	$\log n + 1$	$n + n\log n + \lceil \log n \rceil + 2$
Kogge-Stone	$8n\log n - 14n + 20$	$n\log n - 1$	2	$\log n + 2$	$3n\log n - \frac{n}{2} + 6$
Han-Carlson	$2n\log n - 4\lfloor \frac{n}{2} \rfloor$	$\frac{n\log n}{2}$	2	$\log n + 2$	$\frac{3}{2}n + n\log n - \lfloor \frac{n}{2} \rfloor + 3$
Ladner-Fisher	$3n - 4 + n\log n - 4\lfloor \frac{n}{2} \rfloor$	$\frac{3n}{4} - 1 + \frac{n\log n}{4}$	2	$\log n + 2$	$3n + \frac{n\log n}{2} - \lfloor \frac{n}{2} \rfloor + 1$

3.3.4.2 Comparative Analysis: Optimal Depth Adder vs. Existing Quantum Adders

In this subsection, we assess our proposed optimal depth adder against existing quantum adder designs, as detailed in Table 3.10.

The proposed optimal Toffoli-Depth adder achieves the lowest Toffoli-Depth in the quantum computing, with a harmonious balance in Toffoli Count and qubit count. Specifically, our optimal depth adder with strategy 2 significantly reduces the computational time by halving the Toffoli-Depth to $\log n + 1$, a roughly 50% improvement over the best existing quantum adders. Following closely is our optimal adder with strategy 1, achieving a competitive Toffoli-Depth of $2\log n + 1$.

Interestingly, Fig. 3.36 illustrates that the quantum adders with the lowest Toffoli-Depth prior to our work, specifically the Draper In-place and Out-of-place CLAs [46] and the Quantum Ling adder [226], demonstrate a higher Toffoli-Depth compared to our designs. Nevertheless, despite these existing work have slightly lower Toffoli-Count and Qubit-Count compared to our adders, they do not compensate for the longer time cost associated with their higher Toffoli-Depth.

Overall, our designs substantially enhance Toffoli-Depth, while wisely managing quantum resources, thereby representing a considerable enhancement in both the efficiency and practicality over current quantum adders.

Table 3.10 Performance analysis of different quantum adders. The formula for $\omega(n)$ is $\omega(n) = n - \sum_{y=1}^{\infty} \left\lfloor \frac{n}{2^y} \right\rfloor$ and r represents the radix, with a range of $2 < r \leq n$

Adder	Year	Toffoli count	Toffoli depth	Qubit count
VBE RCA [218]	1995	$4n - 2$	$4n - 2$	$3n + 1$
Cuccaro RCA [40]	2004	$2n - 1$	$2n - 1$	$2n + 2$
Draper in-place CLA [46]	2004	$10n - 3\omega(n) - 3\omega(n-1) - 3\lfloor \log n \rfloor - 3\lfloor \log(n-1) \rfloor - 7$	$8 + \lfloor \log n \rfloor + \lfloor \log(n-1) \rfloor + \left\lfloor \log \frac{n}{3} \right\rfloor + \left\lfloor \log \frac{n-1}{3} \right\rfloor$	$4n - \omega(n) - \lfloor \log n \rfloor$
Draper out-of-place CLA [46]	2004	$5n - 3\omega(n) - 3\lfloor \log n \rfloor - 1$	$4 + \lfloor \log n \rfloor + \left\lfloor \log \frac{n}{3} \right\rfloor$	$4n + 1 - \omega(n) - \lfloor \log n \rfloor$
Takahashi adder [207]	2008	$28n$	$30 \log n$	$2n + \frac{3n}{\log n}$
Takahashi RCA [208]	2009	$2n - 1$	$2n - 1$	$2n + 1$
Takahashi combination [208]	2009	$7n$	$18 \log n$	$2n + \frac{3 \cdot n}{\log n}$
Wang RCAGidney RCA [62]	2018	$2n - 2$	n	$3n - 1$
Gayathri RCA [180]	2021	n	n	$3n + 1$
Higher radix adder [224]	2023	$8n - \frac{n}{r} - (n-1) \pmod{r} - 3\omega(\frac{n}{r}) - 3\log n + 3\log r - 3$	$4\log n + 3r - 2\log r - 2\log 3r + 2\log(r-2) + 2$	$4n - \log n + \frac{n}{r} - \omega(\frac{n}{r}) + \log r - 1$
Quantum Ling adder [226]	2023	$13n - 6\omega(\frac{n}{2}) - 6\lfloor \log \frac{n}{2} \rfloor - 14$	$9 + 2\lfloor \log \frac{n}{2} \rfloor + 2\lfloor \log \frac{n}{6} \rfloor$	$12n - 6\omega(\frac{n}{2}) - 6\lfloor \log \frac{n}{2} \rfloor - 10$
Optimal depth adder+ strategy 1	–	$\frac{3}{2}n \log n + 2\lceil \log n \rceil - n$	$2\log n + 1$	$n + n\log n + \lceil \log n \rceil + 2$
Optimal depth adder+ Strategy 2	–	$\frac{n \log n}{2}$	$\log n + 1$	$n + n\log n + \lceil \log n \rceil + 2$

3.3.4.3 Extension 1: Ling-Based Optimal Toffoli-Depth Adder

Inspired by the innovative quantum Ling adder proposed in Sect. 3.2, we introduce a novel quantum Ling structure based on the prefix tree adders proposed in this work. In Fig. 3.37, a Ling-based adder using the quantum Kogge-Stone is presented as an example, which introduces two parallel prefix computation trees based on the quantum Kogge-Stone computation tree.

This design significantly reduces the Toffoli-Depth of the quantum Ling adder by half. However, this improvement is accompanied by an acceptable increase in Toffoli-Count and qubit count. Specifically, the Toffoli-Depth of our Ling-based Kogge-Stone adder is $2 \log \frac{n}{2} + 8$, reflecting innovative adaptations in generation

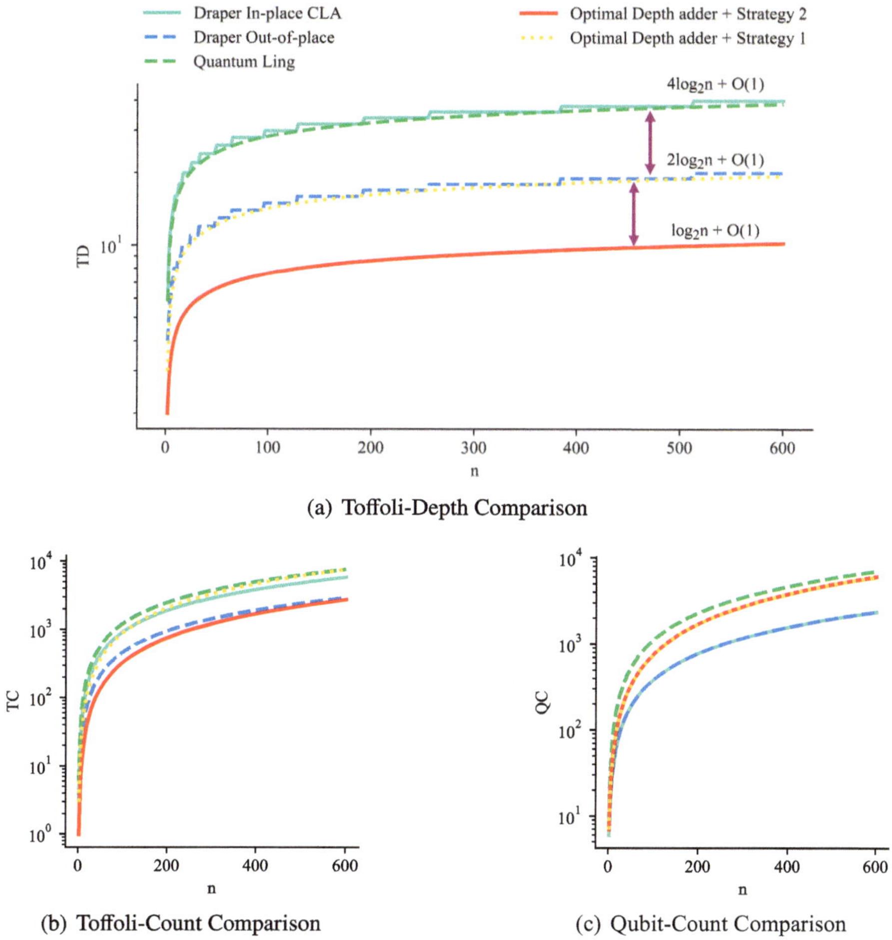

Fig. 3.36 Comparative cost analysis of quantum optimal Toffoli-Depth adder and other top 3 prominent quantum CLA adders

and propagation computations, thereby significantly optimizing the depth of the initial quantum Ling adder [226]. This proposed extension structure demonstrates commendable performance in enhancing the efficiency of Ling-based quantum adders, thereby confirming the flexibility and portability of the adders proposed in this study.

However, as shown in Tables 3.11 and 3.12, we also find that replacing traditional propagation and generation with Ling base is not beneficial for the prefix tree adders we propose. This is because the Ling structure introduces OR logic, requiring two Toffoli gates at each calculation node for propagation and generation. Overall, using Ling base almost doubles the Toffoli-Depth of the prefix tree adders.

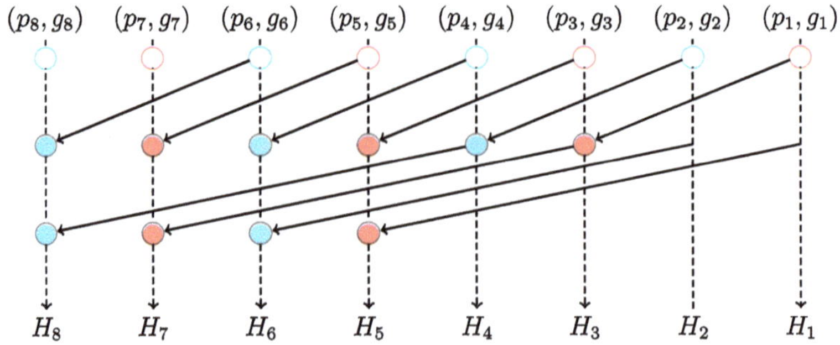

Fig. 3.37 Kogge-Stone prefix tree using Ling expansion

Table 3.11 Performance analysis of Ling-based quantum Kogge-Stone adders utilizing Strategy 1

Adder	Toffoli count	Toffoli depth	Qubit count
Quantum Ling adder [226]	$13n - 6\omega(\frac{n}{2}) - 6\lfloor \log \frac{n}{2} \rfloor - 14$	$9 + 2\lfloor \log \frac{n}{2} \rfloor + 2\lfloor \log \frac{n}{6} \rfloor$	$12n - 6\omega(\frac{n}{2}) - 6\lfloor \log \frac{n}{2} \rfloor - 10$
K-S	$3n \log n + n \log \frac{n}{2} - 3n + 5$	$2 \log n + 2$	$3n \log n - \frac{n}{2} + 6$
K-S+ Ling	$3n \log n + n$	$4 \log n + 6$	$3n \log n + 2n \log \frac{n}{2} + \frac{n}{2} + 3$

Table 3.12 Performance analysis of Ling-based quantum Kogge-Stone adders utilizing Strategy 2

Adder	Extra T count	Toffoli count	Extra T-depth	Toffoli depth	Qubit count
K-S	$8n \log n - 14n + 20$	$n \log n - 1$	2	$\log n + 2$	$3n \log n - \frac{n}{2} + 6$
K-S+Ling	$4n \log n - 8n + 8$	$2n \log n - 4n + 4$	2	$2 \log n + 8$	$3n \log n + 2n \log \frac{n}{2} + \frac{n}{2} + 3$

3.3.4.4 Extension 2: Optimal Toffoli-Depth Modular Adder

In this extension, we utilized the VBE modular addition framework mentioned in the paper [218], which incorporates the quantum prefix tree adders proposed here, as illustrated in Fig. 3.38.

Our quantum modular addition framework primarily contains two important parts: the quantum prefix tree adder and the quantum prefix tree subtractor, with the latter being an altered version of the former. As illustrated in Figs. 3.39a and b, these two structures are very similar to each other.

The implementation of the quantum subtractor is based on certain modifications to the quantum adder, as indicated by Eq. (3.32). Specifically, we begin to take the bit-wise complement of a, and then add b to obtain the result $a' + b$. Finally, we take the two's complement of the qubits of a and the output qubits. Remarkably,

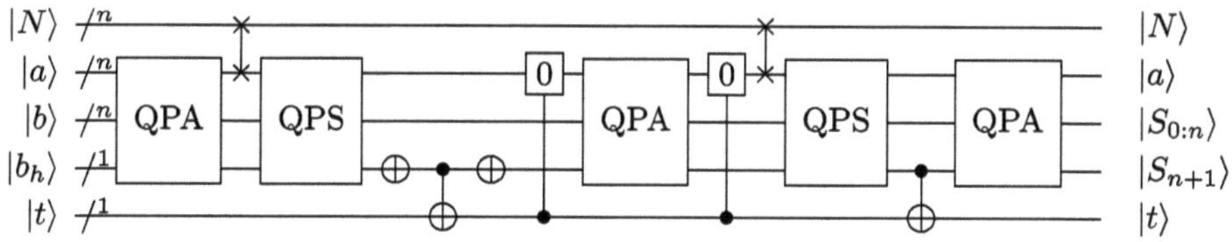

Fig. 3.38 Quantum VBE modular addition framework [218]. Besides, the "Set 0" gate in the framework can be implemented using several CNOTs

(a) Quantum Prefix Tree Adder

(b) Quantum Prefix Tree Subtractor

Fig. 3.39 Two important submodules of the proposed modular addition framework

this process can be easily accomplished using a series of quantum NOT gates. Hence, the cost of the quantum prefix tree subtractor is very close to the cost of the corresponding adder.

$$A - B = (A' + B)'. \tag{3.32}$$

In Table 3.13, we demonstrate the performance of our designed modular adders compared to other existing quantum modular adders. Our designs exhibit superior Toffoli-Depth, particularly the quantum Sklansky with Strategy 2 design achieving an optimal Toffoli-Depth of $5 \log n + 5$. However, this achievement comes at the cost of a higher qubit count and Toffoli-Count.

Above all, our design achieves optimal depth in the quantum world, maintaining an exceptionally low Toffoli-Depth. This accomplishment provides a crucial foundation for efficient arithmetic in quantum computing, unlocking new possibilities for high-performance quantum algorithms and enhancing the overall scalability and practicality of quantum computing.

3.3.5 Section Summary

In conclusion, this work has presented a novel architecture for enhancing the efficiency of quantum adders. By integrating the different prefix tree into the quantum domain, we have achieved a groundbreaking optimal Toffoli-Depth quantum adder, which marks a parallel to the development of classical adder analysis. Furthermore, our work includes a significant quantum Ling expansion, which substantially improves the quantum Ling adder's performance by using our innovative structure.

Table 3.13 Performance analysis of different VBE-based quantum modular adders. The formula for $\omega(n)$ is $\omega(n) = n - \sum_{y=1}^{\infty} \left\lfloor \frac{n}{2^y} \right\rfloor$, where n denotes the bit width of the addend

Modular adder	Year	Toffoli count	Toffoli depth	Qubit count
VBE [218]	1995	$20n - 10$	$20n - 10$	$4n + 2$
Cucarro [40]	2004	$10n - 5$	$10n - 5$	$3n + 3$
Draper in-place [46]	2004	$50n - 15\omega(n) - 15\omega(n-1) - 15\lfloor \log n \rfloor - 15\lfloor \log(n-1) \rfloor - 35$	$40 + 5\lfloor \log n \rfloor + 5\lfloor \log(n-1) \rfloor + 5\left\lfloor \log \frac{n}{3} \right\rfloor + 5\left\lfloor \log \frac{n-1}{3} \right\rfloor$	$5n - \omega(n) - \lfloor \log n \rfloor + 1$
Brent-Kung+Strategy 1	–	$25n - 15\omega(n) - 15\lfloor \log n \rfloor - 5$	$20 + 5\lfloor \log n \rfloor + 5\left\lfloor \log \frac{n}{3} \right\rfloor$	$5n + 2 - \omega(n) - \lfloor \log n \rfloor$
Sklansky+Strategy 1	–	$\frac{15}{2} n \log n + 10\lceil \log n \rceil - 5n$	$10 \log n + 5$	$2n + n \log n + \lceil \log n \rceil + 3$
Kogge-Stone+Strategy 1	–	$15n \log n + 5n \log \frac{n}{2} - 15n + 25$	$10 \log n + 10$	$3n \log n + \frac{n}{2} + 7$
Han-Carlson+Strategy 1	–	$5n + \frac{15}{2} n \log n - 10\left\lfloor \frac{n}{2} \right\rfloor$	$10 \log n + 15$	$\frac{5}{2} n + n \log n - \left\lfloor \frac{n}{2} \right\rfloor + 4$
Ladner-Fisher+Strategy 1	–	$\frac{65n}{4} + \frac{15n \log n}{4} - 10\left\lfloor \frac{n}{2} \right\rfloor - 15$	$10 \log n + 15$	$4n + \frac{n \log n}{2} - \left\lfloor \frac{n}{2} \right\rfloor + 2$
Brent-Kung+Strategy 2	–	$10n - 5 \log n - 10$	$10 \log n - 5$	$5n + 2 - \omega(n) - \lfloor \log n \rfloor$
Sklansky+Strategy 2	–	$\frac{5n \log n}{2}$	$5 \log n + 5$	$2n + n \log n + \lceil \log n \rceil + 3$
Kogge-Stone+Strategy 2	-	$5n \log n - 5$	$5 \log n + 10$	$3n \log n + \frac{n}{2} + 7$
Han-Carlson +Strategy 2	–	$\frac{5n \log n}{2}$	$5 \log n + 10$	$\frac{5}{2} n + n \log n - \left\lfloor \frac{n}{2} \right\rfloor + 4$
Ladner-Fisher+Strategy 2	–	$\frac{15n}{4} - 5 + \frac{5n \log n}{4}$	$5 \log n + 10$	$4n + \frac{n \log n}{2} - \left\lfloor \frac{n}{2} \right\rfloor + 2$

Moreover, our work also incorporates a modular addition extension. The proposed quantum optimal depth adder achieves a Toffoli-Depth of $\log n + O(1)$, which represents a significant improvement over all the previous quantum adders, which had a Toffoli-Depth of at least $2 \log n + O(1)$. Our work marks a significant advancement in the field of quantum computing, addressing the challenge of depth optimization in the area of quantum addition.

3.4 Modular Addition: Quantum Carry-Save Modular Adder

This work aims to explore novel designs for enhancing the efficiency of quantum modular addition. In particular, we introduce a novel quantum modular addition framework based on carry-save architecture, which facilitates the conversion of multiple 2-addend quantum operations within modular addition into a single 3-addend operation, thereby reducing the computational depth. Compared to the most efficient existing quantum modular addition, our design has achieved an impressive result—a reduction in Toffoli Depth by up to 33.33%, while maintaining comparable Toffoli Count and qubit count. This research underscores the potential of carry-save architecture as a promising technique for accelerating quantum modular arithmetic as well as advancing the development of quantum computing in general.

3.4.1 *Introduction*

Shor's algorithm [193] has firmly established itself as an approach in quantum algorithms to efficiently factor large integers into their prime factors. By leveraging on its two key operations, modular exponentiation and its component modular addition, Shor's algorithm can work exponentially faster than classical algorithms. Therefore numerous efforts have been made to push the boundaries on modular exponentiation, as it is one of the most computationally intensive components in Shor's algorithm.

Recent works have achieved many improvements in reducing the bottleneck of modular exponentiation by enhancing the design of its main component, modular addition, through efficient implementations of Clifford+T gates. However, it still remains challenging to develop modular addition frameworks with a lower gate depth and gate count.

In this work, we propose a novel framework for quantum modular addition that relies on the Clifford+T gate set to leverage classical carry-save adders (CSAs). We consolidate multiple 2-addend quantum operations within modular addition into a single 3-addend operation, greatly reducing the overall depth of the quantum circuit.

Our work addresses a hot-in-discussion arithmetic problem: quantum modular addition, expressed as $(a + b) \bmod N$. Here, N represents an n-bit binary number, while a and b represent n-bit binary numbers less than N. This operation usually involves checking if $a+b > N$, and if so, subtracting N to ensure the result is within the range $[0, N - 1]$. In our specific design, we implement the modular addition by involving a three-operand addition $a + b + (-N)$ in the framework.

This study has the following contributions:

- Highlight the potential of carry-save architecture by enhancing the performance of quantum arithmetic designs.

- Reduce the depth of quantum modular addition by proposing a novel framework based on carry-save architecture.
- Accelerate Shor's algorithm through improving modular addition efficiency.

3.4.2 *Existing Modular Addition Frameworks*

Current research on relevant designs can be categorized into the following two main sorts:

3.4.2.1 Clifford+T-Based Designs

Typical architectures realize modular addition with gates in the Clifford+T set. We denote the elementary adder block and subtractor block by *Add* and *Sub* separately. The circuits input quantum states that correspond to the binary representation of the numbers and output $|S\rangle$ as the result of modular addition. Here we enumerate four types of popularly cited modular addition frameworks in literature:

- *VBE Modular Addition Framework.*
  Quantum modular adder based on ripple-carry adders was proposed by Vedral, Barenco, and Ekert in 1996, which we refer to as VBE in this paper [218]. As shown in Fig. 3.40, VBE framework contains five adders and subtractors with reversible CNOT gates and Toffoli gates. After that, Cucarro et al. [40] proposed an improved modular adder with more efficient full adder blocks. However, this framework only solves the modulo calculation when N is the power of two, causing a lack of extendability to general problems. So we omit the Cucarro modular design here.
- *Beckman Modular Addition Framework.*
  Figure 3.41 depicts Beckman's quantum modular framework made from adder-based comparators, full adders, and subtractors [14]. To standardize the performance estimation, we transform all adder-based operations, especially the comparators, into quantum adders and subtractors. Inspired by Beckman's work, Takahashi and Kunihiro propose their modular addition framework composed of two comparators and one adder or subtractor [206]. But their implementation of comparators is based on Draper's Fourier basis addition [45], which implies a

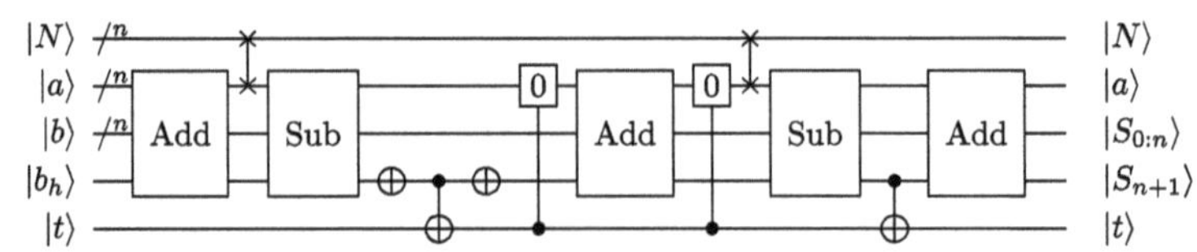

Fig. 3.40 VBE modular addition framework [218]

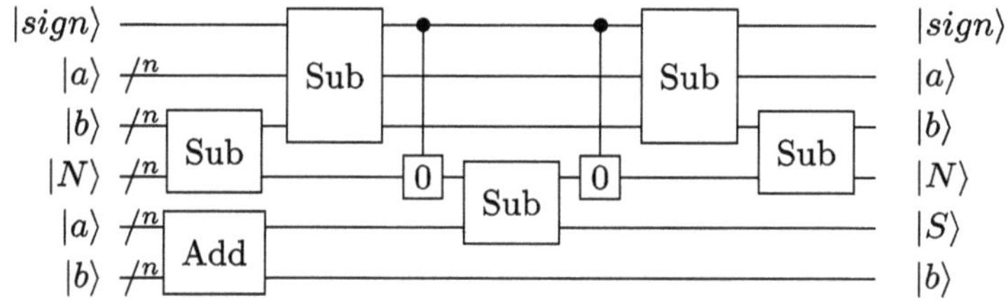

Fig. 3.41 Beckman modular addition framework [14]

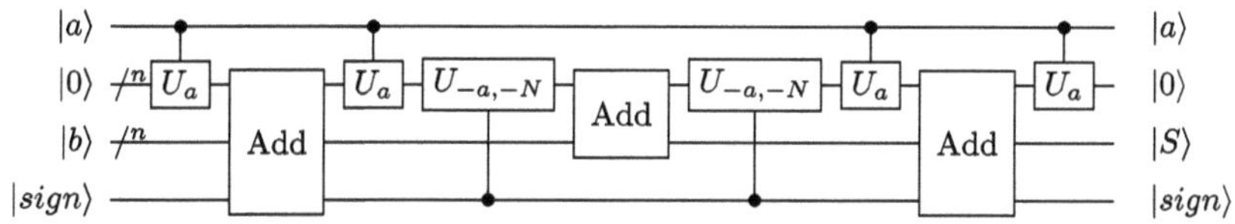

Fig. 3.42 Van Meter modular addition framework [217]

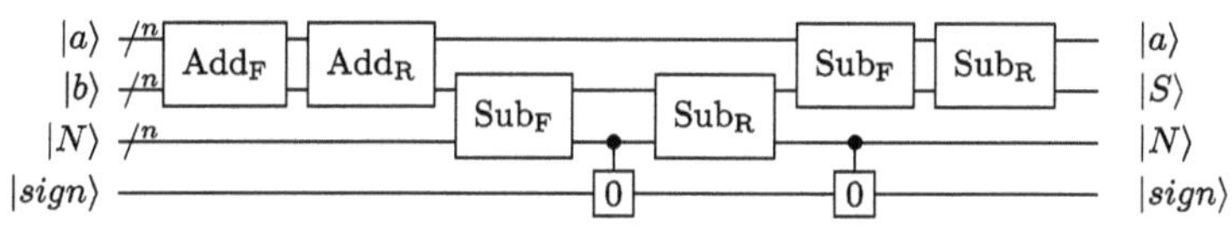

Fig. 3.43 Rines-Chuang modular addition framework [175]

large number of T gates. Haner et al. [83] utilize recursive adder and Carry gates to build the comparators. The intuition is straightforward such that, by comparing the values of b and $N - a$, we can determine whether to add a or subtract $N - a$ and revert the indicator to the zero state.

- *Van Meter Modular Addition Framework.*
  In 2005, Van Meter and Itoh constructed a more efficient modular addition framework by removing the last 2 adder blocks in VBE that were used to undo the carry overflow [217], as is shown in Fig. 3.42. Controlled U_a and $U_{-a,N}$ represent the data loading oracles that set the register with respect to the value of the control qubit. The limitation of this framework is that it requires multiple calls of data loading oracle which complicates circuit operations.
- *Rines-Chuang Modular Addition Framework.*
  In 2018, Rine et al. [175] enhanced Beckman's design [14] by splitting each integer adder into forward and reverse subsections. The implementation of Rines-Chuang's modular addition framework is depicted in Fig. 3.43. It is constructed with one integer adder and two integer subtractors. Every forward and reverse block pairs are denoted with the subscripts F and R, respectively. Intermediate results of the forward block are utilized to control its subsequent reverse block in the second integer adder.

3.4.2.2 QFT-Based Modular Addition Framework

Apart from the circuit designs with Clifford+T gates, modular addition frameworks are also developed in Fourier basis [13, 54, 165]. These algorithms utilize quantum Fourier transformation (QFT) and its inverse as subroutines [9, 243]. However, the complexity associated with QFT arithmetic contributes to its limited adoption in experimental studies. Specifically, the controlled rotation gates with arbitrary angles are practically infeasible to operate in the current stage of hardware than the Toffoli gate. Besides, even the state-of-the-art approximation method to build a QFT block requires $O(n \log n)$ T gates, complicating the experimental realizations [144]. Hence, we only focus on quantum designs based on the Clifford+T gate set in this work.

3.4.3 Carry-Save Architecture

We introduce the carry-save architecture and consider its key implementations.

3.4.3.1 Carry-Save Structure

Various types of adders are available in classical arithmetic, such as ripple-carry adders, carry look-ahead adders, etc. Among these designs, the carry-save adder (CSA) [48] emerges as a high-speed, low-cost design with low latency, making it ideal for efficiently calculating multi-operand additions.

The carry-save architecture is summarized in Fig. 3.44. It achieves the addition of three n-bit inputs by first calculating the partial sum S and shifted carry-out vector C outputs using a carry-save layer. Subsequently, the partial sum and shifted carry-out vector can be added using any n-bit adder, such as a ripple carry adder or carry look-ahead adder, to produce the sum of the multi-operand addition.

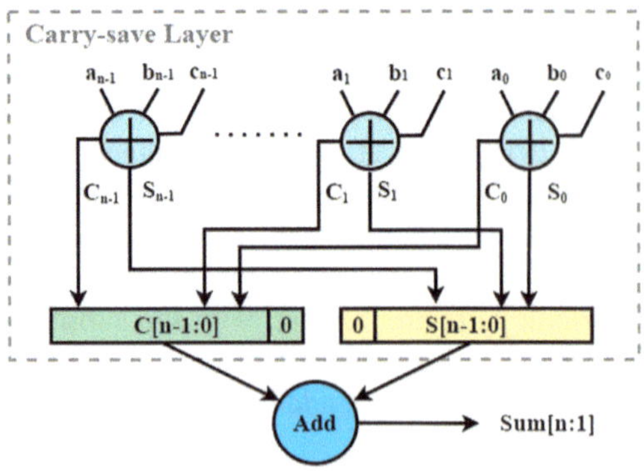

Fig. 3.44 A depiction of the carry-save architecture for multi-operand addition, illustrating the case where the number of operands is 3

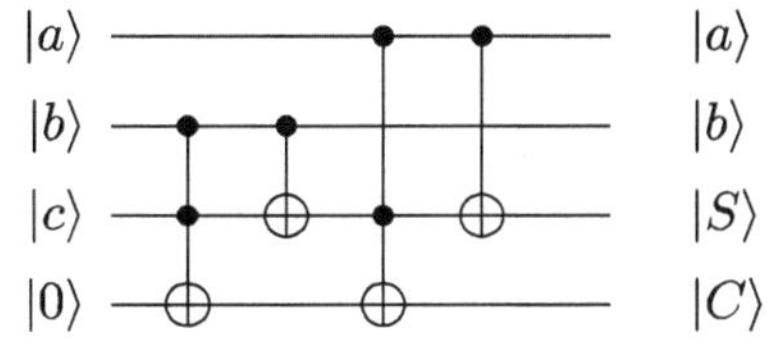

Fig. 3.45 Quantum full adder

3.4.3.2 Quantum Implementation

The quantum implementation of carry-save architecture relies on two primary components: the carry-save layer comprising a series of full adders operating in parallel, and a quantum full adder. Extensive research has already been done on the quantum full adder design [224, 226, 229, 230]. As such, our work will focus on the carry-save layer.

$$S = a \oplus b \oplus c \tag{3.33}$$

$$C = ab \oplus ac \oplus bc. \tag{3.34}$$

We are particularly interested in the shifted carry-out vector C and partial sum S of a 1-bit CSA that can be computed using Formulas (3.33) and (3.34). Particularly, it is worth noting that for hardware implementation, the quantum XOR logic is generally cheaper than OR logic. We note that the function of this basic unit aligns with that of a quantum full adder. Thus by leveraging the quantum full adder circuit illustrated in Fig. 3.45, the quantum carry-save layer can be constructed.

3.4.3.3 Advantages in Efficiency

Previous research [14, 217, 218], suggests that quantum implementations of addition involving three numbers using only 2-addend additions will lead to multiple sequential addition operations. The resulting circuit experiences the full propagation delay of both full adders, giving a depth of $O(2 \cdot TD_{Add})$, where TD_{Add} denotes the Toffoli-Depth of a specific quantum adder for n-bit addition.

In comparison, our quantum implementation uses a carry-save layer and one full adder. The quantum CSA layer only has a Toffoli-Depth 2. Its resulting CSA-based quantum circuit for 3-addend addition will possess a Toffoli-Depth of only $O(2 + TD_{Add})$, which significantly improves the sequential addition implementation.

3.4.4 Methodology

Our proposed quantum carry-save-based modular addition framework is summarized in Fig. 3.48. It contains six operations. Below, we describe some of the novel components and structure of our framework.

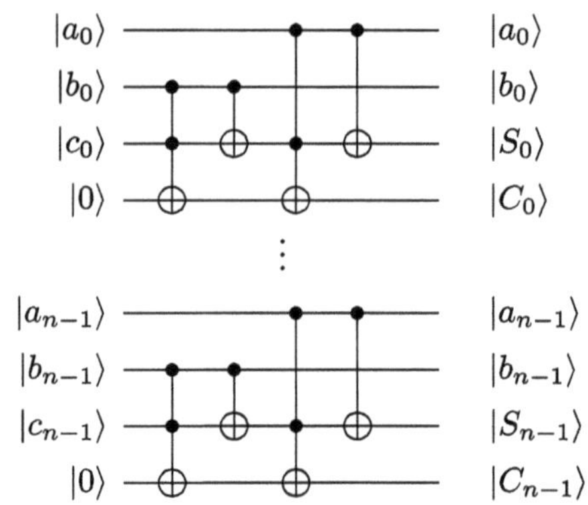

Fig. 3.46 Quantum carry-save architecture

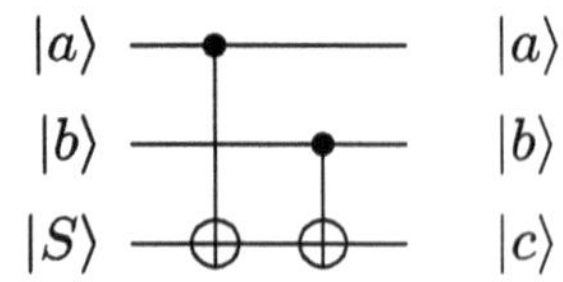

Fig. 3.47 Quantum uncompute S module

3.4.4.1 Main Components

The following are the five main components of our design:

- **CSA Module** represents the carry-save architecture utilized in our work. As demonstrated in Sect. 3.4.3, the CSA module consists of a series of full adders operating in parallel, as depicted in Fig. 3.46.
- **Add Module** represents a quantum adder. In quantum arithmetic, there are various types of adders, such as ripple-carry adders [40, 218], carry look-ahead adders [46, 224, 226, 232], etc. In this chapter, we adopt prominent works detailed in Sect. 3.4.5.
- **Submodule** denotes a quantum subtractor, which is a direct application of the quantum adder.

$$a - b = (a' + b)'. \tag{3.35}$$

  As shown in Eq. (3.35), the process involves bit-wise complementation of a, followed by its addition to b, yielding $a' + b$. Subsequently, the bits of a and the output qubits are complemented. Given that bit-wise complementation can be achieved using quantum NOT gates, the cost of a quantum subtractor is close to that of the corresponding adder.
- **Set 0 Module** offers the same as the VBE Modular framework [218], and the set 0 operation can be implemented using a series of CNOT gates.
- **Uncompute S module** is used to uncompute the partial sum S, which can be achieved through a series of CNOT gates, as detailed in Fig. 3.47.

3.4.4.2 Overall Structure

- **Step 1**: The carry-save layer is performed to calculate the partial sum, S, and the shifted carry, C. For n-bit modular addition, all inputs and outputs in this step have a length of n qubits. In terms of the detailed costs, the TC is $2n$, the QC is $4n$, and the TD is 2.
- **Step 2**: A signed addition is applied to S and C to calculate the sum of $(a+b-N)$, with the initial sign set to 0. The associated costs include a TC of TC_{Add}, a TD of TD_{Add}, and there is an additional ancilla qubit required, totaling $1 + Anc_{\text{Add}}$.
- **Step 3**: Utilize Set 0 Module, which is constructed only with CNOT gates. The sign qubit serves as the control qubit for this module. If the sign is in the state $|0\rangle$, indicating $(a+b-N) \geq 0$, the Set 0 gate toggles $|-N\rangle$ to the state $|0\rangle$. Otherwise, $|-N\rangle$ passes through the module unchanged. This step incurs no requirement for TC, TD, and additional ancilla.
- **Step 4**: Perform a subtraction if the sign qubit is $|0\rangle$, computing $(a+b-N)-0$. Otherwise, $(a+b-N)-(-N)$ is calculated to restore the original value of $a+b$. As for the cost, TC is TC_{Add}, and TD is TD_{Add}, with no additional ancilla required.
- **Step 5**: The previous Set 0 Module is uncomputed by applying all the CNOT gates in reverse order. For this step, there is no additional TC, TD, or ancilla required.
- **Step 6**: After applying the Uncompute S Module in the final step, all qubits except the final result of $(a+b) \mod N$ are uncomputed. Here, no additional TC, TD, or ancilla are required.

$$\begin{aligned}\text{TD} = &\overbrace{2}^{\text{Step 1}} + \overbrace{TD_{Add}}^{\text{Step 2}} + \overbrace{0}^{\text{Step 3}} + \overbrace{TD_{Add}}^{\text{Step 4}} + \overbrace{0}^{\text{Step 5}} + \overbrace{0}^{\text{Step 6}} \\ =&2 + 2 \times TD_{Add}\end{aligned} \tag{3.36}$$

$$\begin{aligned}\text{TC} = &\overbrace{2n}^{\text{Step 1}} + \overbrace{TC_{add}}^{\text{Step 2}} + \overbrace{0}^{\text{Step 3}} + \overbrace{TC_{add}}^{\text{Step 4}} + \overbrace{0}^{\text{Step 5}} + \overbrace{0}^{\text{Step 6}} \\ =&2n + 2 \times TC_{Add}\end{aligned} \tag{3.37}$$

$$\begin{aligned}\text{QC} = &\overbrace{4n}^{\text{Step 1}} + \overbrace{1 + Anc_{Add}}^{\text{Step 2}} + \overbrace{0}^{\text{Step 3}} + \overbrace{0}^{\text{Step 4}} + \overbrace{0}^{\text{Step 5}} + \overbrace{0}^{\text{Step 6}} \\ =&4n + 1 + Anc_{Add}.\end{aligned} \tag{3.38}$$

We present a comprehensive construction for a carry-save-based quantum modular addition framework in these six steps. The detailed costs are illustrated in Formulas (3.36), (3.37), and (3.38), where TC_{Add} represents the Toffoli-Count,

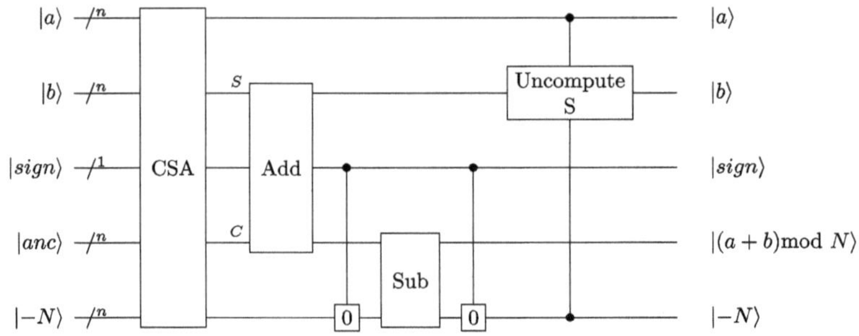

Fig. 3.48 Our modular addition framework

Table 3.14 Efficient quantum adders utilized in Experiment 1

Adder	TD	TC	QC
Gayathri [180]	n	n	$3n + 1$
Takahashi [208]	$2n - 1$	$2n - 1$	$2n + 1$
Optimal TD [232]	$\log n + 1$	$\frac{n \log n}{2}$	$n + n \log n + \lceil \log n \rceil + 2$

Table 3.15 Performance analysis of different quantum modular addition frameworks

Framework	Year	Toffoli-Depth	Toffoli-Count	Qubit count
VBE [218]	1996	$5 \times TD_{Add}$	$5 \times TC_{Add}$	$3n + 2 + Anc_{Add}$
Beckman [14]	1996	$5 \times TD_{Add}$	$6 \times TC_{Add}$	$5n + 1 + 4 \times Anc_{Add}$
Van Meter [217]	2005	$3 \times TD_{Add}$	$3 \times TC_{Add}$	$3n + 1 + Anc_{Add}$
Rines-Chuang [175]	2018	$3 \times TD_{Add}$	$3 \times TC_{Add}$	$3n + 1 + 2 \times Anc_{Add}$
Our design	–	$2 + 2 \times TD_{Add}$	$2n + 2 \times TC_{Add}$	$4n + 1 + Anc_{Add}$

TD_{Add} represents the Toffoli-Depth, and Anc_{Add} denotes the ancilla count of the n-bit quantum add building block (as shown in Fig. 3.48).

3.4.5 Results and Discussion

3.4.5.1 Experiment 1: Comparison with Prominent Quantum Modular Addition Frameworks.

After reviewing Paper [232], we choose three adders in Table 3.14 as our adder blocks since they exhibit optimal TC, TD, QC among all other quantum adders, respectively. Table 3.15 clearly demonstrates that the effectiveness of a proposed quantum modular addition framework largely relies on the add module, representing a quantum adder. A performance comparison between our proposed quantum modular addition framework and other relevant prominent works is summarized in Fig. 3.49, where we note the following:

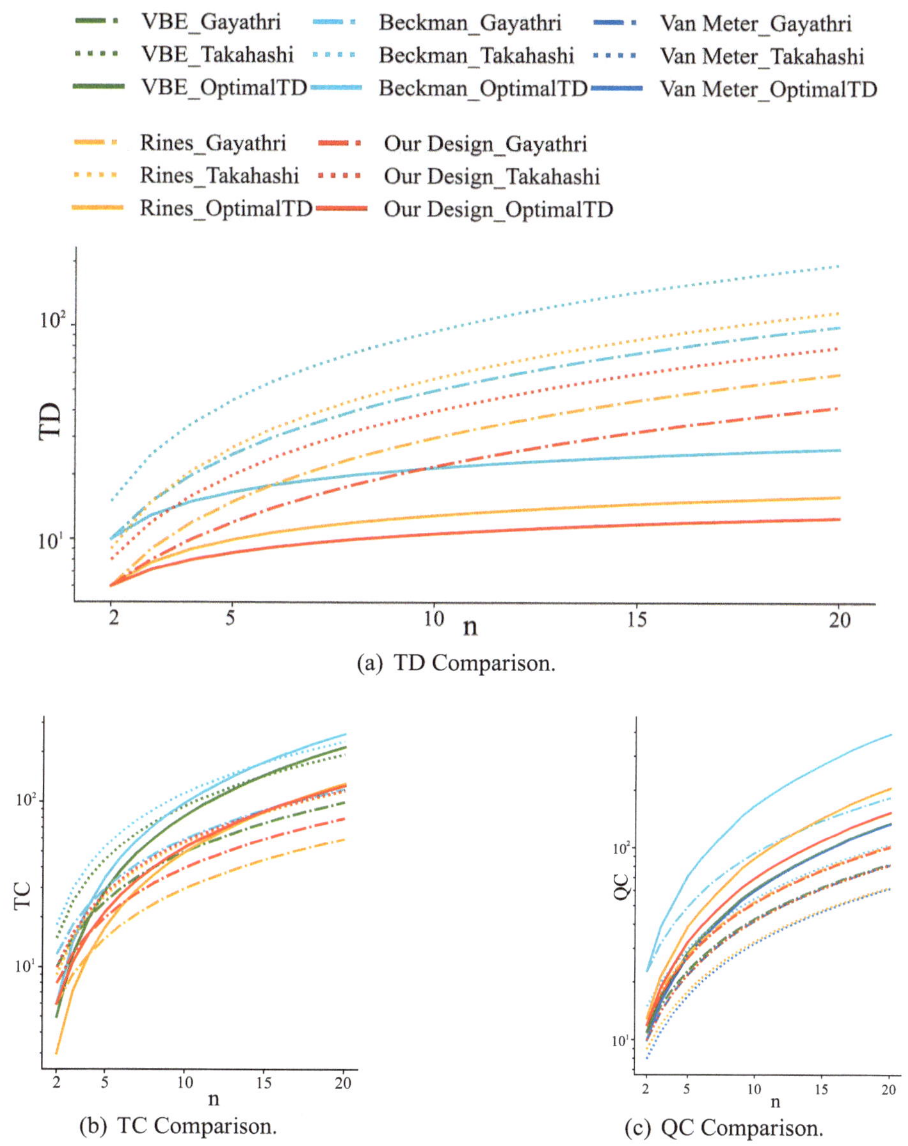

Fig. 3.49 Comparative cost analysis of the proposed quantum modular adders. Here, n denotes the qubit width of the modular addition

- The framework structure determines the modular adder efficiency. Our structure consistently demonstrates the lowest TD compared to all others, as shown in Fig. 3.49a. In terms of TC, the Rines-Chuang framework with Gayathri Add block achieves the lowest TC cost, with our design closely following as the second lowest, as depicted in Fig. 3.49b. However, when using a low TC add

block, the Rines-Chuang framework gains more significant advantages. Our design exhibits a QC that falls slightly below the Beckman framework, as shown in Fig. 3.49c. When the ancilla of the Add block is high, our design shows similar QC to the Rines-Chuang framework.
- The choice of Add Block significantly impacts the overall performance of modular adders. Within the same modular framework, utilizing Gayathri Adder achieves the lowest TC, while employing Optimal TD Adder results in the lowest TD. Besides, using Takahashi Adder allows for achieving the lowest QC.
- Our Optimal TD Add block design minimizes notable time requirements by achieving the lowest TD, comparing our Gayathri Add block variation, which scores the second lowest TC and shows its advantage in reducing Toffoli resource usage. On the other hand, our design incorporating the Takahashi Add block achieves the fourth lowest QC.

Our designs exhibit commendable efficiency across various cost metrics in comparison to existing efficient quantum modular addition framework. We achieved up to 33.33% reduction in TD, while keeping TC and QC at comparable levels.

3.4.5.2 Experiment 2: Influence on Quantum Modular Exponentiation

We leverage the VBE modular exponentiation framework [218] to analyze the influence of our proposed modular addition framework on quantum modular exponentiation circuits. We start with constructing an n-bit controlled multiplier to compute $ax \bmod N$, where x is represented as an n-bit binary number. This task is accomplished by using repeated modular N adders. The corresponding TD, TC, and QC cost of this circuit are detailed in Formulas (3.39), (3.40), and (3.41), respectively. Additionally, it is important to note that the cost of the set state module within this circuit depends on the specific scenario. Under the best case, TC and TD costs are 0, whereas under less favorable circumstances, TC and TD costs are n.

$$TD_{C-MODMul} = 2n \times SETState + n \times TD_{MODAdd} \tag{3.39}$$

$$TC_{C-MODMul} = 2n \times SETState + n \times TC_{MODAdd} \tag{3.40}$$

$$QC_{C-MODMul} = 1 + n + QC_{MODAdd}. \tag{3.41}$$

Building upon the n-bit controlled modular multiplier, the VBE modular exponentiation can be constructed to calculate $a^x \bmod N$, where x is an n-bit binary number. Formulas (3.42), (3.43), and (3.44) illustrate the number of calls to our modular adder block required to perform an n-bit modular exponentiation.

$$\begin{aligned} TD_{MODExp} &= 2n \times TD_{C-MODMul} \\ &= 4n^2 SETState + 2n^2 TD_{MODAdd} \end{aligned} \tag{3.42}$$

$$\begin{aligned} TC_{MODExp} =& 2n \times TC_{C-MODMul} \\ =& 4n^2 SETState + 2n^2 TC_{MODAdd} \end{aligned} \tag{3.43}$$

$$\begin{aligned} QC_{MODExp} =& n + QC_{C-MODMul} \\ =& 1 + 2n + QC_{MODAdd}. \end{aligned} \tag{3.44}$$

By employing our proposed structure, a 60% reduction in addition operations in both VBE quantum modular multiplication and VBE quantum modular exponentiation can be achieved.

3.4.6 Section Summary

In this research, we focused on exploring innovative designs to enhance the efficiency of quantum modular addition. We presented a novel quantum modular addition framework based on carry-save architecture, reducing the computational intensity associated with modular addition. When compared against existing quantum modular addition framework, our framework can reduce up to 33.33% in TD, while maintaining comparable TC and QC.

3.5 Division: Quantum Non-restoring Divider

In this study [230], we focus on enhancing the performance of quantum slow dividers by exploring the design choices of its subblocks, such as adders. Through comprehensive design space exploration of state-of-the-art quantum addition building blocks, our work has resulted in an impressive achievement: a reduction in Toffoli-Depth of up to 93.90%, accompanied by substantial reductions in both Toffoli- and qubit count of up to 92.12% and 99.38%, respectively. This study offers crucial perspectives on efficient design of quantum dividers and emphasizes the importance of adopting a systematic design space exploration approach.

3.5.1 Introduction

As a fundamental arithmetic operation in both classical and quantum computing, the development of division is crucial in important quantum algorithms such as computing shifted quadratic character problems [81], principal ideal problems [214], and hidden shift problems [215]. In the past, significant effort was spent to design and optimize division for classical computing [58, 59, 210]. In quantum computing, this problem is relatively unexplored. As we will show in this work, there is significant room for improving the performance. Moreover, we emphasize the importance for a systematic design space exploration for building large-scale designs.

Design space exploration (DSE) is a standard practice for designing large circuits in classical computing [135]. However, it is not yet applied to the quantum circuit designs of quantum division. We begin by decomposing non-restoring division into various submodules, prominently consisting of adders. Since there are many quantum adders offering diverse trade-offs in logical depth and size (gate count, qubit count), the natural question arises about the selection of a specific quantum adder. We also ensure that the proposed circuit is based on the Clifford+T gate set. This is crucial in permitting fault-tolerant and scalable quantum computation. This study makes the following substantial contributions:

- Comprehensive application of DSE methodology during quantum division circuit implementation
- Achievement of unprecedented performance in quantum division circuits through careful consideration of all relevant metrics

3.5.2 *Methodology*

3.5.2.1 Quantum Non-restoring Division

In this section, we describe the construction workflow of our addition-based non-restoring divider, divided into quotient generation phase and remainder restoration phase. The detailed pseudocode and quantum circuit are presented in Algorithm 1 and Fig. 3.50, respectively.

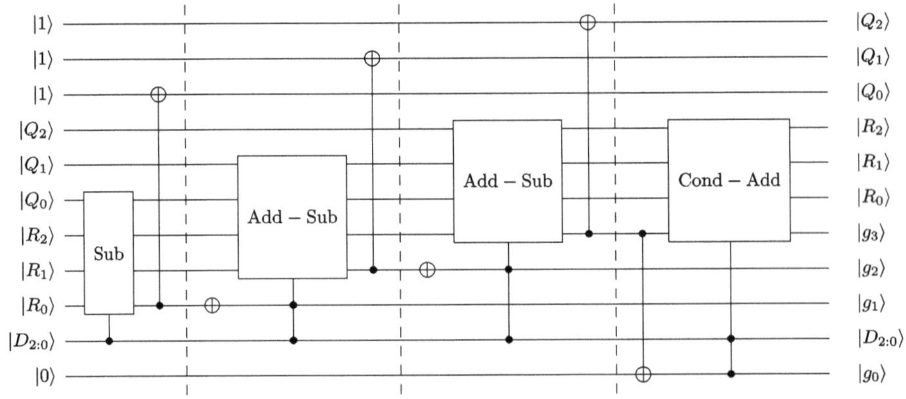

Fig. 3.50 3-qubit quantum non-restoring divider

Algorithm 1 Non-restoring division (Q,D)

Require: Non-zero unsigned integers Q represents the dividend and D represents the divisor.
Q and D are two n bit input values.
Returns qubits Q as quotient, and qubits R as remainder.

```
R = 0^{n+1}        ▷ Where 0^{n+1} are n + 1 zero bits and R and Q are qubit pairs.
Q = Q_{n-1} Q_{n-2} ··· Q_1 Q_0
D = 0 D_{n-1} D_{n-2} ··· D_1 D_0
/* Quotient Generation Phase Start */
for i = 0 to n do
    if R_0 = 0 then
        (R, Q) = 2 * (R, Q)        ▷ Left-shift operation
        R = R − d                  ▷ Subtraction operation
    else
        (R, Q) = 2 * (R, Q)
        R = R + d                  ▷ Addition operation
    if R_0 = 0 then                ▷ Set Q_{n-1} based on R_0
        Q_{n-1} = 1
    else
        Q_{n-1} = 0
/* Quotient Generation Phase End */
/* Remainder Restoration Phase Start */
if R_0 = 1 then                    ▷ Check for valid remainder
    R = R + d
/* Remainder Restoration Phase End */
Return Q, R
End
```

For our design, the essential sub-circuits include the subtractor, controlled adder-subtractor, and conditional adder. As shown in Fig. 3.51a, the subtractor sub-circuit effectively transforms quantum adders into subtractors with inputs $|a\rangle$ and $|b\rangle$. Specifically, the subtraction $|a\rangle - |b\rangle = |a\rangle + |\bar{b}\rangle + 1$ is achieved by flipping the subtrahend bits and setting the carry-in to high. In Fig. 3.51b, we present the implementation of our controlled adder-subtractor sub-circuit. The input bit $|R_{i-1}\rangle$, where i represents the current iteration in the quotient generation phase, functions as both the control and carry-in qubit. When $|R_{i-1}\rangle$ is equal to $|1\rangle$, it flips the subtrahend qubits for subtraction, performing addition otherwise. Furthermore, our conditional adder module is illustrated in Fig. 3.51c based on previous implementations [210]. When the control qubit is in the high state, this circuit produces the sum $|s\rangle = |a\rangle + |b\rangle$. Otherwise, $|a\rangle$ and $|b\rangle$ pass through the circuit unchanged.

In the optimization of our workflow, we observe that the first iteration of the quotient generation phase always involves subtraction due to the initialized constant value (0) of qubit `R` post-initialization. Besides, the left-shift operation of the qubit pair (`R`, `Q`) can be omitted by appropriately offsetting the sub-circuits of subsequent operations. These optimizations result in substantial conservation of quantum resources: (i) reduction in the width of `R`, (ii) partial execution of the first left-shift operation, and (iii) simplifying the initial arithmetic sub-circuit into a quantum subtractor.

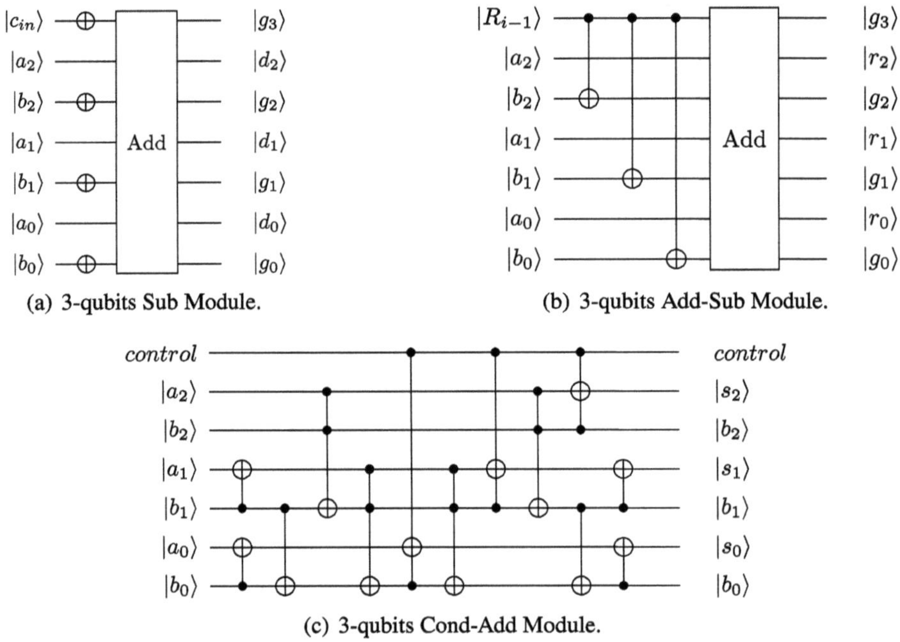

Fig. 3.51 The three essential sub-circuits in the proposed quantum divider

$$
\begin{aligned}
\mathrm{TD} &= \overbrace{TD_{add}}^{\text{Step 1}} + \overbrace{(TD_{add}+1)\cdot(n-1)}^{\text{Step 2}} + \overbrace{3n-2}^{\text{Step 3}} \\
&= n\cdot TD_{add} + 4n - 3
\end{aligned}
\tag{3.45}
$$

$$
\begin{aligned}
\mathrm{TC} &= \overbrace{TC_{add}}^{\text{Step 1}} + \overbrace{(TC_{add}+3)\cdot(n-1)}^{\text{Step 2}} + \overbrace{3n-2}^{\text{Step 3}} \\
&= n\cdot TC_{add} + 6n - 5
\end{aligned}
\tag{3.46}
$$

$$
\begin{aligned}
\mathrm{QC} &= \overbrace{3n+2+Anc_{add}}^{\text{Step 1}} + \overbrace{n-1}^{\text{Step 2}} + \overbrace{1}^{\text{Step 3}} \\
&= 4n + 2 + Anc_{add}.
\end{aligned}
\tag{3.47}
$$

As shown above, Eqs. (3.45), (3.46), and (3.47) show the overall cost of the proposed quantum circuit for an n-qubit division. This analysis is divided into three steps: Step 1 corresponds to the initial iteration of the quotient generation phase, Step 2 presents the subsequent iterations of the quotient generation phase, and Step 3 shows the remainder restoration phase.

Our design is presented at a high-level model, affording flexibility in the choice of the quantum adder. Hence, TC_{add} and TD_{add} denote the Toffoli-Count and Toffoli-Depth of the selected adder, respectively. Moreover, the variable Anc_{add} represents the ancilla qubits required by the chosen adder.

3.5.2.2 An Extension Design: Quantum Restoring Divider

As an extension of our research, we propose an addition-based restoring quantum divider, drawing from the workflow in the previous section. For n-qubit division, the corresponding TD, TC, and QC are formulated as $(n \cdot TD_{add} + 3n^2 - 2n)$, $(n \cdot TC_{add} + 3n^2 - 2n)$, and $(4n + 1 + Anc_{add})$, respectively.

As shown in Table 3.17, with a carefully selected addition block, our non-restoring design consistently outperforms restoring divider in both TD and TC, with the addition of only one ancilla. Therefore, we only focus on the non-restoring divider for its superior overall efficiency. A detailed description of this extension will be accessible online upon this work's acceptance.

3.5.3 Results and Discussions

In this section, we provide detailed results and discussions.

3.5.3.1 Impact of Quantum Adders on Quantum Dividers

As shown in Table 3.16, the selection of the adder building block has a substantial impact on the performance of our divider.

Here, we systematically evaluated various building blocks based on TC, TD, and QC to identify the most suitable choices for different scenarios:

- **Toffoli-Depth.** Toffoli-Depth significantly influences the execution time of the quantum circuit. As shown in Table 3.16, CLA adders usually achieve low Toffoli-Depth by using the parallel prefix tree approach to maximize parallel computation subblocks and reduce calculation depth. For example, using Draper in-place CLA as the basic building block achieves superior Toffoli-Depth compared to other quantum RCAs and combination adders such as Takahashi RCA and Takahashi combination adder. Significantly, quantum Ling adder-based non-restoring divider achieves a substantial reduction in Toffoli-Depth complexity from $O(n \log(n))$ to $O(n \log(\frac{n}{2}))$.
- **Toffoli-Count.** When the primary target is to minimize the Toffoli-Count, RCA-based addition building blocks offer a great solution. As shown in Fig. 3.52b and Table 3.16, these RCA-based building blocks are characterized by their simplicity and low Toffoli-Count, making them an attractive choice for Toffoli-

Table 3.16 Performance of the proposed n-qubit quantum non-restoring divider based on different addition building blocks. The equation for $\omega(n)$ is $\omega(n) = n - \sum_{y=1}^{\infty} \lfloor \frac{n}{2^y} \rfloor$, and r represents the radix, with a range of $2 < r \leq n$

Quantum division circuit			
	Performance		
Constituent adder circuit	Toffoli-Depth	Toffoli-Count	Qubit count
Variable/Unspecified adder	$n \cdot TD_{add} + 4n - 3$	$n \cdot TC_{add} + 6n - 5$	$4n + 2 + Anc_{add}$
VBE RCA [218]	$4n^2 + 2n - 3$	$4n^2 + 4n - 5$	$5n + 3$
Cuccaro RCA [40]	$2n^2 + 3n - 3$	$2n^2 + 5n - 5$	$4n + 4$
Draper in-place CLA [46]	$12n + n \lfloor \log n \rfloor + n \lfloor \log(n-1) \rfloor + n \lfloor \log \frac{n}{3} \rfloor + n \left\lfloor \log \frac{n-1}{3} \right\rfloor - 3$	$10n^2 - 3n \cdot \omega(n) - 3n \cdot \omega(n-1) - 3n \lfloor \log n \rfloor - 3n \lfloor \log(n-1) \rfloor - n - 5$	$6n - \omega(n) - \lfloor \log n \rfloor + 2$
Takahashi low-ancilla adder [207]	$30n \log n + 4n - 3$	$28n^2 + 6n - 5$	$4n + \frac{3n}{\log n} + 2$
Takahashi RCA [208]	$2n^2 + 3n - 3$	$2n^2 + 5n - 5$	$4n + 3$
Takahashi combination [208]	$18n \log n + 4n - 3$	$7n^2 + 6n - 5$	$4n + \frac{3n}{\log n} + 2$
Wang RCA[223]	$n^2 + 4n - 3$	$n^2 + 6n - 5$	$5n + 3$
Gidney RCA[62]	$n^2 + 4n - 3$	$2n^2 + 4n - 5$	$5n + 1$
Gayathri RCA[180]	$n^2 + 4n - 3$	$n^2 + 6n - 5$	$5n + 3$
Higher radix adder [224]	$4n \log n + 3r \cdot n - 2n \log r - 2n \log 3r + 2n \log(r-2) + 6n - 3$	$8n^2 - \frac{n^2}{r} - n \cdot (n-1) \pmod r - 3n \cdot \omega(\frac{n}{r}) - 3n \log n + 3n \log r + 3n - 5$	$6n - \log n + \frac{n}{r} - \omega(\frac{n}{r}) + \log r + 1$
Quantum Ling adder [226]	$13n + 2n \lfloor \log \frac{n}{2} \rfloor + 2n \lfloor \log \frac{n}{6} \rfloor - 3$	$13n^2 - 6n \cdot \omega(\frac{n}{2}) - 6n \lfloor \log \frac{n}{2} \rfloor - 8n - 5$	$14n - 6\omega(\frac{n}{2}) - 6 \lfloor \log \frac{n}{2} \rfloor - 8$

Count-sensitive scenarios. Among all the building blocks, it is noteworthy that the Gayathri RCA and Wang RCA achieving the lowest cost in terms of Toffoli-Count by using only n Toffoli gates to complete n-qubits addition.

- **Qubit-Count.** In quantum computing, evaluating the qubit count is crucial for efficient quantum resource management. Although CLA-based designs achieve low Toffoli-Depth, they usually require more ancilla qubits. As illustrated in Fig. 3.52c, RCA-based building blocks clearly demonstrate their advantages in achieving competitive qubit count. Apart from the RCA-based building blocks, an appealing alternative is Takahashi's combination adder [208], which strategically integrates components from both RCA and CLA, resulting in a low qubit count that scales linearly with the division bit width n. Remarkably, Takahashi's low-ancilla adder [207] has the same qubit count as Takahashi's combination adder [208], making it an equally appealing choice. Therefore, RCA-based adders [207] and [208] are highly recommended as the preferred building blocks in QC cost-dominated situations.

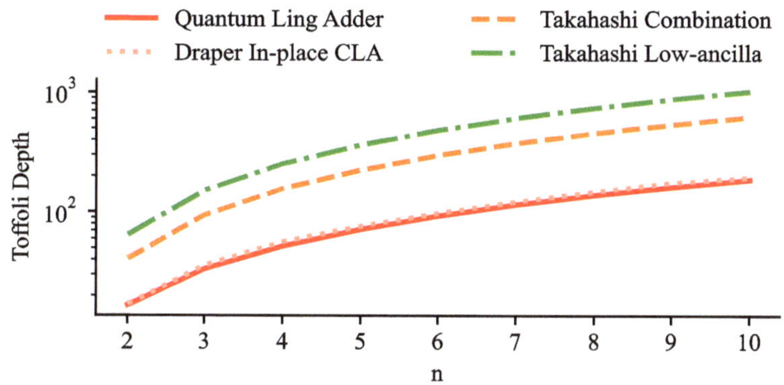

(a) Low Toffoli-Depth Non-restoring Divider.

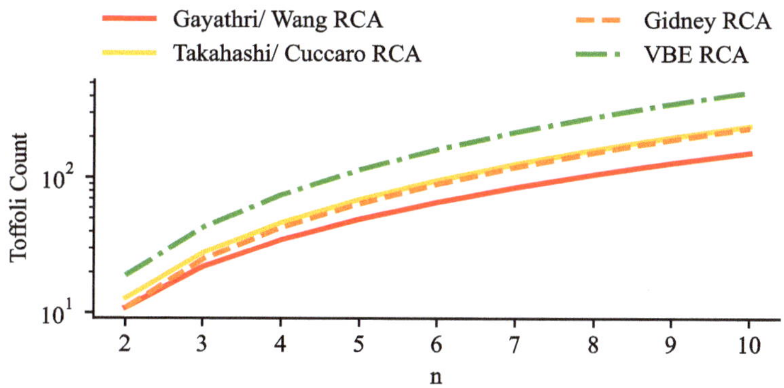

(b) Low Toffoli-Count Non-restoring Divider.

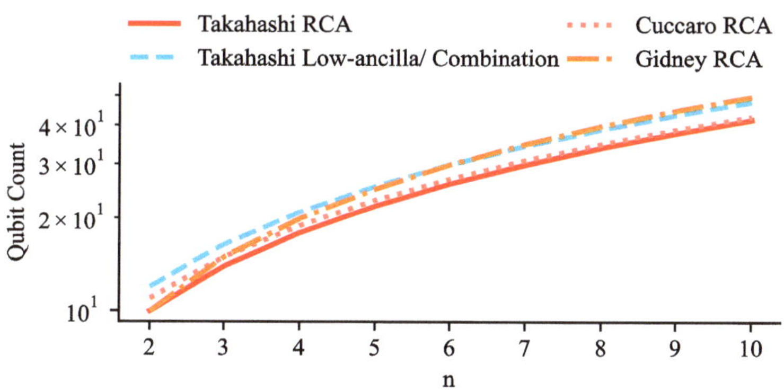

(c) Low Qubit-Count Non-restoring Divider.

Fig. 3.52 The proposed quantum non-restoring dividers based on different adders. Here, n denotes the qubit width of the division. Given the identical performance in Toffoli-Count between Gayathri RCA and Wang RCA, we use a simplified notation "Gayathri/Wang RCA" to represent their performance. The similar notations are also applied for "Takahashi/Cuccaro RCA" and "Takahashi Low-ancilla /Combination"

Table 3.17 Comparison of proposed 32-qubit quantum dividers with existing designs. In this table, "Takahashi C" refers to the Takahashi combination adder. Percentage improvement is calculated w.r.t. the quantum Newton-Raphson divider [58]

Divider	Toffoli-Depth	Toffoli-Count	Qubit count
Goldschmidt [59]	17, 850	117, 187	30, 008
Newton-Raphson [58]	13, 506	93, 376	23, 996
Restoring+Ling	3, 707 (72.55%)	12, 032 (87.11%)	319 (98.67%)
Restoring+ Takahashi C	5, 888 (56.40%)	10, 176 (89.10%)	149 (99.38%)
Non-restoring+Ling	824 (93.90%)	9, 211 (90.14%)	320 (98.67%)
Non-restoring+ Takahashi C	3, 005 (77.75%)	7, 355 (92.12%)	150 (99.37%)

Overall, as illustrated in Figs. 3.52a, b, and c, the most effective addition building blocks for minimizing Toffoli-Depth, Toffoli-Count, and qubit count are identified as quantum Ling adder, Gayathri/Wang RCA, and Takahashi RCA, respectively.

3.5.3.2 Comparison with Existing Quantum Dividers

Following a comprehensive design space exploration to assess the impact of quantum adders on our design, we constructed and compared the proposed restoring and non-restoring dividers with existing work. The quantum Ling adder, which has minimal Toffoli-Depth, achieves 93.90% in Toffoli-Depth savings, and the Takahashi combination adder, which offers balanced results across all metrics, sees 92.12% and 99.37% savings in Toffoli-Count and qubit count, respectively.

The specific data for 32-qubit division is shown in Table 3.17. Obviously, our design outperforms all the prior designs in terms of TD, TC, and QC, owing to the absence of resource-intensive conditional statements and operations in the proposed workflows.

3.5.4 *Section Summary*

In conclusion, this study explored the efficiency of quantum non-restoring divider, focusing on the impact of integrating different quantum adders. This required an in-depth study of the division algorithm, expressing it in terms of addition circuits. We observed that when minimizing the Toffoli-Depth is crucial, a divider based on quantum CLA adders offers Toffoli-Depth reductions of up to 93.90%. When minimizing Toffoli-Count is essential, the quantum RCAs are a more apt choice, achieving reductions up to 92.12%. Lastly, both RCA-based adders and the Takahashi combination adder offer optimal solutions for prioritizing Qubit-Count at up to 99.38% improvement.

3.6 Overview: Quantum Arithmetic Circuits

In this overview, we aim to provide a systematically organized and easily comprehensible overview of the current state of the art in quantum arithmetic circuits. Specifically, this study covers fundamental operations such as addition, subtraction, multiplication, division, and modular exponentiation. We delve into the detailed quantum implementations of these prominent designs and evaluate their efficiency considering various objectives.

3.6.1 Introduction

With the continuous advancement of quantum computers, the importance of quantum arithmetic circuits is increasing. Consequently, substantial research efforts are focused on studying more efficient arithmetic circuits and improving quantum-based techniques to develop these critical units, meeting application requirements in the quantum field.

In this domain, several surveys have been conducted. For instance, Takahashi [204] summarizes efficient quantum circuits for basic arithmetic operations used in Shor's factoring algorithm, focusing on quantum addition, comparison, and the quantum Fourier transform for addition. Additionally, F. Orts et al. [156] have conducted a detailed investigation of one of the most essential quantum arithmetic circuits—the quantum reversible adder. Moreover, Sousa [198] extensively reviewed various nonconventional computer arithmetic approaches, including quantum arithmetic circuits.

This work aims to provide a detailed survey of the latest developments in basic quantum arithmetic circuits by examining two mainstream designs: those based on the Clifford+T set and those based on the QFT. Compared to previous studies, our work offers a more comprehensive coverage of recent advancements in various basic arithmetic operations, providing detailed technical background and implementation details. Through this work, we aim to provide researchers and engineers with a thorough reference to better understand and apply the latest quantum arithmetic designs, thereby advancing the field of quantum computing.

3.6.1.1 Contribution

In this chapter, we provide a comprehensive review of the latest designs in quantum arithmetic circuits, delving into various performance metrics and data formats. Quantum arithmetic circuits primarily rely on two mainstream foundational methods: one based on Clifford+T gates and the other on QFT. Our work thoroughly explores both of these primary design approaches in various aspects, as detailed in Fig. 3.53.

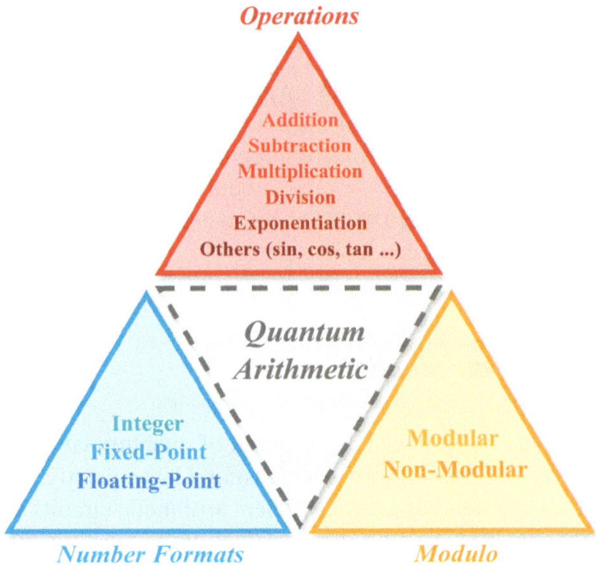

Fig. 3.53 Three main aspects of quantum arithmetic designs discussed in this overview

3.6.2 *Quantum Addition*

Addition is a fundamental operation that serves as a cornerstone for the functionality of all algorithms built upon it. In classical computing, addition is typically performed by adders such as ripple-carry, carry look-ahead [176], carry-save [49], carry-select, and carry-skip. Similarly, Clifford+T-based and QFT-based quantum adders can be classified into these same categories. Here, we focus exclusively on integer addition designs.

Clifford+T-Based Designs

Table 3.18 presents an overview of prominent Clifford+T-based quantum adders, detailing the publication year, corresponding article, structure, and key innovations for each design. Additionally, Fig. 3.54 is provided to illustrate the articles published over the years categorized in the four main structures discussed here for Clifford+T-based quantum adders:

- **Ripple-Carry Structure.**
  The initial design for Clifford+T-based quantum addition dates back to 1996, when Vedral et al. [218] presented an explicit construction of quantum networks capable of performing fundamental arithmetic operations. Among these operations was the initial design of a quantum n-bit ripple-carry adder, marking a significant milestone in the development of quantum arithmetic.
  In 2004, Cucarro et al. [40] introduced a new linear-depth ripple-carry quantum addition circuit, following Vedral et al.'s 1996 proposal of a ripple-carry

Table 3.18 An overview of prominent Clifford+T-based quantum adders

Year	Article	Structure	Key innovations
1996	Vedral et al. [218]	Ripple-carry	Propose the first quantum ripple-carry adder
1998	Gossett [70]	Carry-save	Propose the first quantum carry-save adder
2004	Cuccaro et al. [40]	Ripple-carry	Use only one ancilla
2004	Draper et al. [44]	Carry look-ahead	Propose the first quantum carry look-ahead adder
2005	Takahshi et al. [205]	Ripple-carry	Use zero ancilla
2008	Takahshi et al. [207]	Hybrid	Design ripple-carry & Carry look-ahead adder
2009	Takahshi et al. [208]	Ripple-carry	Reduce the size & Depth of adder with zero ancilla
2009	Takahshi et al. [208]	Hybrid	Simplify the design [207]
2009	Takahshi et al. [208]	Modified hybrid	Simplify the design [208]
2016	Wang et al. [223]	Ripple-carry	Simplify the building blocks of ripple-carry adder
2018	Gidney [63]	Ripple-carry	Introduce logical-AND to reduce the cost
2021	Gayathri et al. [180]	Ripple-carry	Simplify the building blocks of ripple-carry adder
2023	Wang et al. [224]	Hybrid	Simplify the building blocks of hybrid adder
2023	Wang et al. [226]	Carry look-ahead	Propose quantum Ling base adder
2024	Wang et al. [232]	Carry look-ahead	Explore parallel prefix trees in quantum

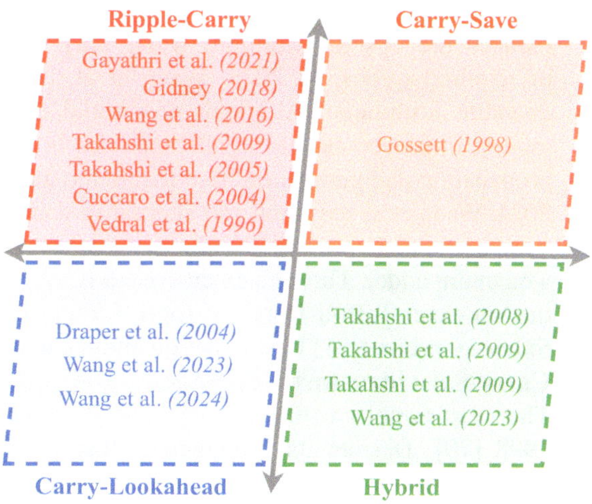

Fig. 3.54 Classification of Clifford+T-based quantum adders based on the four main structures discussed here

adder [218]. While Vedral et al.'s adder [218] required a linear number of ancillary qubits, this newer design [40] relies on just a single ancillary qubit. Additionally, the circuit boasts reduced depth and a smaller gate count compared to earlier designs. In the following year, Takahashi et al. [205] further enhanced the quantum ripple-carry adder by developing a quantum circuit for adding two n-bit binary numbers without any ancillary qubit. In 2009, Takahashi et al. [208] further improved the ripple-carry adder. This design, with a size of $7n - 6$, was smaller than any previous quantum circuit for addition without ancillary qubits. Additionally, Wang et al. [223] introduced an enhanced linear-depth ripple-carry quantum addition circuit. Unlike the previous adders, which typically required at least two Toffoli gates per output bit, this adder utilized only a single Toffoli gate per bit.
In 2018, Gidney [63] introduced a significant design advancement. By devising a novel "logical-AND" construction employing four T gates to record the logical-AND of two qubits into an ancilla and zero T gates to subsequently erase the ancilla, this approach effectively halved the cost of current quantum addition. Moreover, Gayathri et al. [180] improved the structure of the quantum full adder and introduced a quantum ripple-carry adder based on this improvement. To manage the garbage output during the uncomputing phase of the proposed quantum adder, they used a technique involving circuit reversal using the reversible pebble game in accordance to the Bennet removal scheme[16]. As a result, they developed an uncomputing quantum circuit that does not rely on Toffoli gates, thereby reducing the excess number of Toffoli gates in the proposed adder circuit.
- **Carry Look-Ahead Structure.**
After the introduction of quantum ripple-carry adders, Draper et al. [44] introduced an efficient addition circuit inspired by classical carry look-ahead arithmetic circuits. This pioneering quantum carry look-ahead adder significantly reduces the cost of addition, while only marginally increasing the number of required qubits. In 2023, Wang et al. [226] introduced the Ling basis in quantum arithmetic. Although this initial quantum Ling adder showed only modest improvement, its significance lies in exploring the potential to alter the propagation and generation base within the quantum carry look-ahead design. In 2024, Wang et al. explored 160 alternative compositions of the carry-propagation structure in their work [232], aiming to determine the optimal depth structure for a quantum adder. Through extensive study of these structures, they demonstrated that an exact Toffoli-Depth of $logn + O(1)$ is attainable by implementing the Sklansky prefix tree [195] in the quantum carry look-ahead adder.
- **Carry-Save Structure.**
The most famous work on carry-save structures was introduced by Gossett in 1998 [70]. Despite its prominence, this work still offers considerable room for improvement. Additionally, carry-save structures find extensive application in constructing modular multiplication and modular exponentiation arithmetic circuits, as elaborated in Sect. 3.6.6.

Table 3.19 An overview of prominent QFT-based quantum adders

Year	Article	Structure	Key innovations
2000	Draper [45]	Quantum-quantum adder	Propose the first quantum-quantum adder
2002	Beauregard [13]	Quantum-classical adder	Propose the first quantum-classical adder
2012	Pavlidis et al. [165]	Quantum-classical adder	Design multi-controlled adder
2017	Ruiz-Perez et al. [177]	Quantum-quantum adder	Realize non-modular summation
2020	Sahin [254]	Quantum-quantum adder	Perform addition on signed integers
2021	Pavlidis et al. [164]	Quantum-quantum adder	Propose the first QFT-based qudit adder
2022	Seidel et al. [188]	Quantum-quantum adder	Propose adder with semi-Boolean polynomials
2023	Atchade-Adelomou et al. [9]	Quantum-classical adder	Summarize in-place and out-of-place structure
2023	Cakmak et al. [253]	Quantum-quantum adder	Execute the adder in experimental demonstration

- **Hybrid Structure.**
  The ripple-carry approach, while reducing the number of qubits, is known to incur large Toffoli-Depth [40, 218]. Conversely, the carry look-ahead approach decreases Toffoli-Depth but requires a larger qubit count [44]. Hence, in 2008, the first hybrid structure was introduced to leverage the low qubit count and the low Toffoli-Count advantages of the ripple-carry adder as well as the low Toffoli-Depth advantage of the carry look-ahead structure. Specifically, Takahashi et al. [207] combined a modified version of Draper et al.'s quantum carry look-ahead adder [44] with parallel applications of Takahashi et al.'s quantum ripple-carry adder [205]. In 2009, Takahashi et al. [208] proposed a generalized and simplified version of the previous hybrid method [207] for constructing a logarithmic-depth circuit with a reduced number of qubits. Additionally, they introduced a more efficient version of this hybrid design using unbounded fan-out gates. In 2023, Wang et al. [224] introduced an efficient quantum carry look-ahead adder utilizing a higher radix structure. They achieved this by separating the propagation and summation phases within the Manchester carry chain, resulting in a flexible and efficient hybrid addition circuit. This approach allows for various carry paths and sum paths within the quantum higher radix addition framework, enhancing its flexibility and efficiency in different scenarios.

QFT-Based Circuit Designs

Quantum addition circuits with Fourier transform are categorized into two main structures based on the type of input factors. Advancements in QFT-based adders are listed in Table 3.19. To better visualize the timeline of quantum-quantum adders

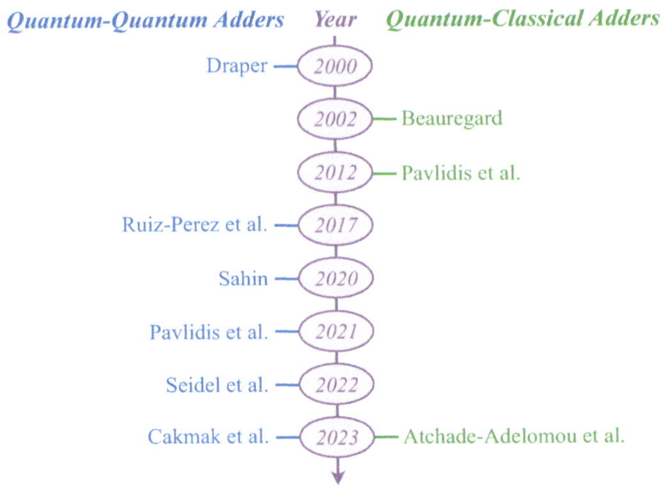

Fig. 3.55 Timeline of QFT-based adders based on structure type

and quantum-classical adders published over the years, a timeline is provided in Fig. 3.55.

- **QFT-Based Quantum-Quantum Adder.**
  Given a and b encoded as two n-qubit quantum states $|a\rangle$ and $|b\rangle$, the addition circuit of $|a\rangle$ and $|b\rangle$ requires at least $2n$ qubits as inputs. One of the factors, for example $|b\rangle$, is used as the ancilla qubits that control the rotation angles in the phase gates. An illustrative diagram is shown in Fig. 3.56, where the controlled phase gates realize the summation of b on the Fourier basis of a. This architecture was first proposed by Draper in 2000 [45].
  Later, Ruiz-Perez and Garcia-Escartin summarized the design for both modular and non-modular operations by introducing an additional carry qubit and used the adders to compute weighted sums [177]. In 2020, Sahin applied Ruiz-Perez and Garcia-Escartin's framework for arithmetic operations on signed integers and presented algorithms for calculating two's complement, absolute value, and comparison [254].
  QFT-based addition was also studied in qudit systems, where data is encoded with a higher dimensional unit [164]. In 2022, Seidel et al. proposed arithmetic for signed and unsigned integers and floating points using a novel data encoding scheme based on semi-boolean polynomials [188]. In their design, integers are encoded as semi-boolean polynomials by a QFT block and decoded by its inverse. Additions with respect to the semi-boolean polynomials are performed using controlled phase gates in the middle of the circuit. Therefore, we categorize this framework as similar to the QFT-based quantum-quantum adder.
  In 2023, Cakmak constructed a primitive quantum arithmetic logic unit capable of performing AND and logical NOT-AND operations and executed it on the IBM quantum platform as an experimental demonstration [253].

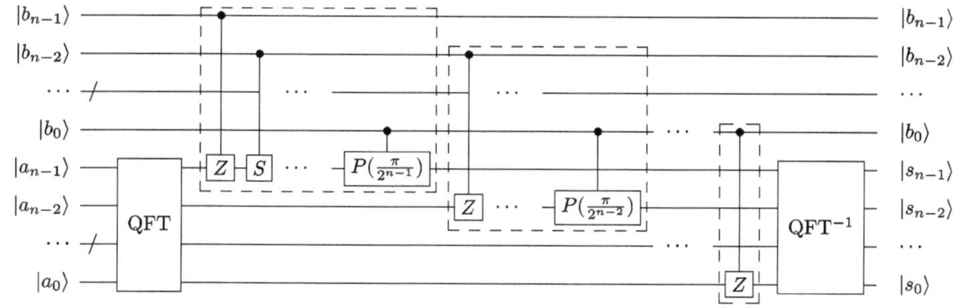

Fig. 3.56 Architecture of QFT-based adder with two quantum states $|a\rangle$ and $|b\rangle$ as inputs

Fig. 3.57 Architecture of QFT-based adder with a quantum state $|a\rangle$ and a classical constant b as inputs

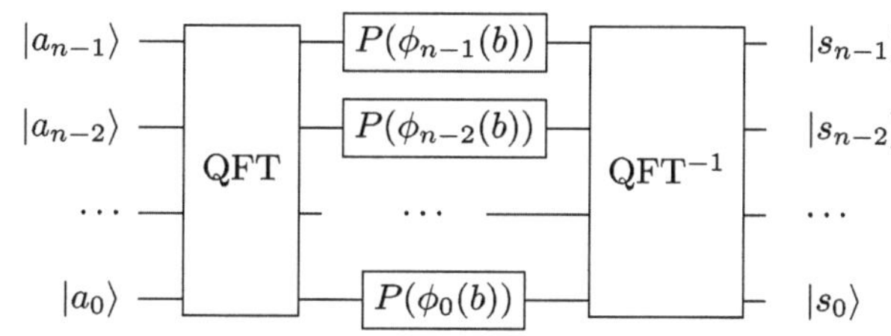

- **QFT-Based Quantum-Classical Adder.**
  Another form of the addition circuit takes the inputs of a quantum state and a classical constant. Given the $|a\rangle$ as an initial state and b as a classical value, the QFT-based adder is constructed by precomputing the rotation angles with respect to b and performing single-qubit rotations accordingly. The circuit has a reduced number of input qubits. Figure 3.57 illustrates the adder, which encodes the classical summation factor b into rotation angles $\phi(b)$ in the phase gates, realizing the summation of b on the Fourier basis of a. This design was first proposed by Beauregard in 2002 as a modified framework of Draper's adder [13]. In 2012, Pavlidis and Gizopoulos proposed the controlled quantum adder and the controlled-controlled quantum adder for arithmetic with a classical constant [165]. As elaborated in the sections below, these two gates are pivotal for high-level arithmetic operations such as multiplication and exponentiation. In 2023, Atchade-Adelomou and Gonzalez further summarized quantum-classical adders with in-place and out-of-place operations, discussing their applications on intermediate-scale platforms [9].

3.6.3 *Quantum Subtraction*

N-bit quantum binary subtractors have structures closely resembling those of quantum adders. We discuss the construction of quantum subtraction circuits for

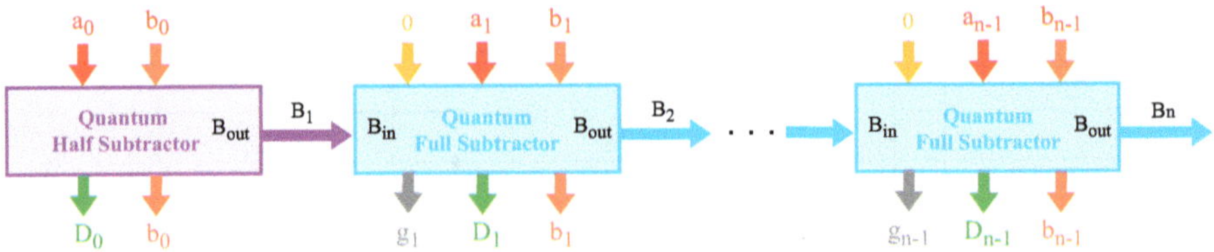

Fig. 3.58 n-bit quantum subtractor based on Method 1

both Clifford+T design and QFT-based designs. These circuits can be categorized into three primary methods, as detailed below.

Method 1: Ripple-Borrow Subtraction

Subtractor circuits, which use two n-bit operands to produce an n-bit result with a borrow-out signal, are less frequently discussed in quantum systems compared to quantum addition circuits. Despite this, quantum subtractor circuits offer significant potential for innovative designs in quantum computing.

The quantum full subtractor module differs only slightly from the quantum full adder module, and similarly, the quantum half subtractor module differs only slightly from the quantum half adder module. By using quantum half subtractors and full subtractors, an n-bit binary quantum subtractor can be constructed using the ripple-borrow approach, as illustrated in Fig. 3.58. Numerous existing quantum works [32, 209, 211] utilize this method to develop their quantum subtractors. These designs share a similar structure and overall cost with ripple-carry adder circuits, typically requiring a linear Toffoli-Depth.

Method 2: Subtraction Based on Addition Without Input Carry

In addition to Method 1, it is possible to construct an n-bit quantum subtractor by making minor modifications to an n-bit quantum adder circuit, thereby using it as a subtractor. These modifications, as exemplified in Formulas (3.48) and (3.49), can be achieved simply by employing quantum NOT gates.

$$A - B = (A' + B)' \tag{3.48}$$

$$A - B = A + B' + 1. \tag{3.49}$$

Method 2, illustrated in Formula (3.48), relies on this principle. Specifically, the design of the quantum n-bit subtractor involves complementing the input a at the outset and subsequently complementing both a and the resulting sum at the end. As shown in diagram 3.59, the proposed subtractor design heavily relies on the structure of the quantum adder used. Notably, these modifications only entail the use of quantum NOT gates, thus eliminating the need for extra Toffoli gates or qubits.

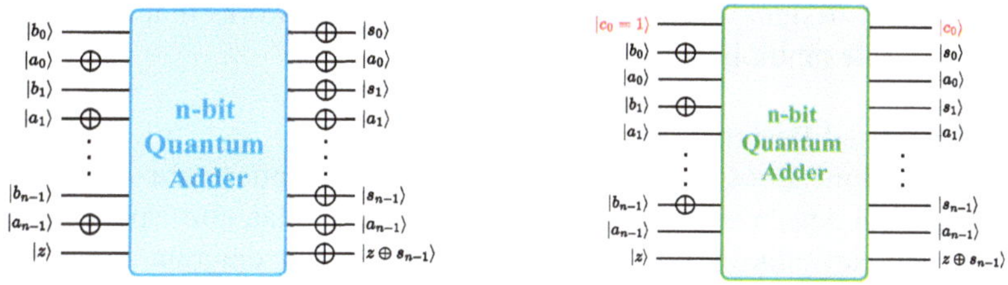

(a) Method 2 (b) Method 3

Fig. 3.59 n-bit quantum subtractor

Table 3.20 Cost comparison between Method 1 and Method 2 for n-bit quantum subtraction

Subtraction	NOT gates	Qubit count	Carry in
Method 2	$3n + 1$	$2n + 1$	✗
Method 3	n	$2n + 2$	✓

For QFT-based designs, Method 2 is the most commonly used principle for constructing a subtractor from an adder. According to the commuting rule of X gate and $P(\phi)$ phase gate, $XP(\phi)X = P(-\phi)$, interpolating quantum NOT gates as a minor modification to the QFT-based adder block is equivalent to a new QFT-based adder block with the same structure but negative rotation angles, as mentioned in Beauregard et al.'s design [13] and Sahin's work [254].

Method 3: Subtraction Based on Addition with Input Carry
In contrast to Method 2, Method 3 leverages an n-bit quantum adder with input carry to construct an n-bit quantum subtractor. As shown in Formula (3.49), this method realizes an n-bit quantum subtractor by employing a quantum adder circuit with input carry, as illustrated in diagram 3.59. Specifically, the design process involves complementing the input b at the outset and setting the input carry c_0 to 1. In brief, this method is similar to Method 2, with slight differences in the number of NOT gates and qubit count, as shown in Table 3.20.

3.6.4 *Quantum Multiplication*

In classical computation, advanced multiplication methods such as Shift-and-Add, Karatsuba [111], Wallace Tree [221], and Toom-Cook [37] algorithms minimize the number of multiplicative operations, thereby boosting efficiency. Fast Fourier transform (FFT)-based methods, such as the Schönhage-Strassen algorithm [184], further improve the efficiency of large integer multiplication by transforming it into a convolution. Similarly, in quantum multiplication, these classical methods are leveraged to perform operations even more efficiently. In this section, we focus solely on binary designs, although there are also several works that concentrate on ternary reversible multipliers [8, 51, 159].

Clifford+T-Based Designs
Quantum multiplications using Clifford+T-based multipliers are categorized in Table 3.21, with details on their respective publication year, corresponding article, structure, and key innovations. A general classification diagram is provided in Fig. 3.60, grouping each article based on its structure and sorted by publication year for quantum Clifford+T-based multipliers.

- **Schoolbook.**
  A simple approach to integer multiplication is to reduce it to addition using n adders. This method, known as the schoolbook method, was applied in both Draper et al. [44] and Parent et al. [160]'s papers. While this early methodology features a straightforward structure, it is not efficient in terms of qubit count, Toffoli-Count, or Toffoli-Depth.
- **Shift-and-Add.**
  Shift-and-add multiplication is similar to the traditional grade-school multiplication algorithm. In the quantum module library QLib, proposed by Lin et al. [127], quantum modules of various sizes and specifications are available for benchmarking quantum logic and physical synthesis. Although the overall cost is high, the multiplier within QLib represents an initial quantum design based on the shift-and-add multiplication method.
  In 2016, Jayashree et al. [108] proposed a quantum multiplier based on a modified version of the shift-and-add method. This design uses reversible logic to create a garbage-free and ancilla qubit-optimized quantum integer multiplier. It addresses the issue of redundant garbage qubits in existing reversible quantum integer multipliers. Specifically, it reduces the required ancilla qubits from $3N + 1$, as in the design by Lin et al. [127], to $2N + 1$ for N-bit quantum multiplication.
  In 2018, Muñoz-Coreas et al. [143] introduced a T-count optimized quantum integer multiplier using the shift-and-add method. Based on a novel quantum conditional adder circuit, this design achieved a T-count optimized quantum circuit, requiring only $4N + 1$ qubits and generating no garbage outputs. This design significantly reduces the Toffoli-Count and the Toffoli-Depth compared to Jayashree et al. [108]'s work, while maintaining the same number of ancilla qubits.
  In 2022, Sajadimanesh et al. [183] extended the applicability of the quantum shift-and-add multiplier to real quantum NISQ devices. They achieved this by employing approximate computing, which leverages the resilience of applications to errors and prioritizes efficiency over precision. By developing approximate multipliers that trade accuracy for reduced complexity, the researchers demonstrated that running the proposed approximate multiplier on real quantum computers produces acceptable noise levels for practical applications.

Table 3.21 An overview of prominent Clifford+T-based quantum multipliers

Year	Article	Structure	Key innovations
1998	Zalka [245]	Schönhage-Strassen	Propose the first Schönhage-Strassen multiplier
2004	Parent et al. [160]	Schoolbook	Use n plain adders in schoolbook method
2004	Draper et al. [44]	Schoolbook	Use n Carry look-ahead adders
2014	Lin et al. [127]	Shift-and-Add	Propose the first shift-and-add multiplier
2015	Kepley et al. [113]	Karatsuba	Propose the first quantum Karatsuba multiplier
2016	Jayashree et al. [108]	Shift-and-Add	Optimize ancillas and eliminate garbage
2017	Parent et al. [160]	Karatsuba	Optimize qubit count for Karatsuba multiplier
2018	Dutta et al. [47]	Toom-Cook 2.5	Propose the first quantum Toom-Cook multiplier
2018	Muñoz-Coreas et al. [143]	Shift-and-Add	Optimize Toffoli-Count and Toffoli-Depth
2019	Gidney [64]	Karatsuba	Use inline-mutation calls to optimize qubit count
2019	Gidney [65]	Windowed	Introduce lookup tables
2020	Van Hoof [216]	Karatsuba	Optimize qubit count for Karatsuba multiplier
2021	Larasati et al. [122]	Toom-Cook 3	Optimize qubit count and Toffoli-Depth
2021	Gayathri et al. [61]	Wallace tree	Propose the first quantum Wallace tree multiplier
2022	Sajadimanesh et al. [183]	Shift-and-Add	Adapt the design for quantum devices
2023	Putranto et al. [168]	Toom-Cook 8	Optimize space & Time cost for Toom-Cook design
2023	Orts et al. [154]	Wallace tree	Support arbitrary size multiplication
2023	Nie et al. [149]	Schönhage-Strassen	Improve Schönhage-Strassen multiplier
2023	Jang et al. [105]	Karatsuba	Reduce the Toffoli-Depth to one

- **Karatsuba.**
  The Karatsuba algorithm [111] is a fast multiplication algorithm that uses a divide-and-conquer approach to multiple 2 n-bit numbers. In 2015, Kepley et al. [113] first used Karatsuba's multiplication algorithm in the quantum domain to develop a software implementation that automatically generates a quantum multiplication circuit. The number of T-gates in this circuit is limited to $7 \cdot n^{\log 3}$ for any given reduction polynomial of degree $n = 2^N$. If an irreducible trinomial of degree n exists, a multiplication circuit with a total gate count of $O(n^{\log 3})$ can

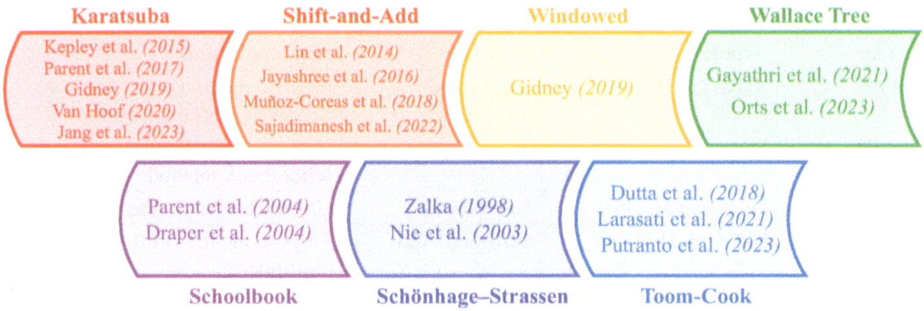

Fig. 3.60 General classification of Clifford+T-based quantum multipliers articles as per structure and sorted by publication year

be achieved. Compared to benchmarks [138] and schoolbook designs [44, 160], this synthesized design uses fewer Toffoli gates and Clifford gates, although it requires more qubits.

Two years later, Parent et al. [160] introduced an enhanced version of the previous quantum Karatsuba-based integer multiplication. These improvements resulted in a marginal increase in operations by a factor of less than 2 and a slight asymptotic rise in depth for the parallel version. The asymptotic enhancements stem from a thorough analysis of pebble games on complete ternary trees. Specifically, the primary improvement over the previous circuit proposed by Kepley et al.'s [113] lies in the asymptotic reduction of required qubits, decreasing from $O(n^{1.585})$ to $O(n^{1.427})$, while maintaining a similar Toffoli-Count and Toffoli-Depth.

In 2019, Gidney [64] improved the space complexity of Parent et al. [160]'s quantum Karatsuba multiplication from $O(n^1.427)$ to $O(n)$, while maintaining similar Toffoli complexity. Gidney achieved this by enabling recursive calls to directly contribute their outputs to subsections of the output register, eliminating the need to store and uncompute intermediate results. This technique, similar to classical tail-call optimization, is crucial for optimizing the space complexity of quantum recursive algorithms and can be included in any quantum algorithm design.

The following year, Van Hoof [216] introduced space-efficient variants of the Karatsuba multiplier, requiring 3n qubits. This approach maintains a Toffoli-Count and Toffoli-Depth similar to Kepley et al. [113]'s work but incurs a higher number of CNOT gates, theoretically up to $O(n^2)$.

In previous studies, the optimization of quantum Karatsuba multiplication has primarily focused on reducing Toffoli-Count or Qubit-Count, neglecting the importance of minimizing Toffoli-Depth. In 2023, Jang et al. [105] aimed to address this gap by prioritizing the reduction of Toffoli-Depth and full depth in quantum multiplication circuits. Their optimized quantum multiplication approach achieves a Toffoli-Depth of one, significantly lowering the overall depth

Table 3.22 Asymptotic performance analysis of prominent Toom-Cook quantum multipliers

Article	Structure	Qubit count	Toffoli-Count	Toffoli-Depth
Putranto et al. [168]	Toom-Cook 2	$O(n^{1.589})$	$O(n^{\log_2 3})$	$O(n^{1.217})$
Dutta et al. [47]	Toom-Cook 2.5	$O(n^{1.404})$	$O(n^{\log_6 16})$	$O(n^{1.143})$
Larasati et al. [122]	Toom-Cook 3	$O(n^{1.35})$	$O(n^2)$	$O(n^{1.112})$
Putranto et al. [168]	Toom-Cook 4	$O(n^{1.313})$	$O(n^{\log_4 7})$	$O(n^{1.09})$
Putranto et al. [168]	Toom-Cook 8	$O(n^{1.245})$	$O(n^{\log_8 15})$	$O(n^{1.0569})$

of the quantum circuit. While their design requires more qubits, it has considerably lower Toffoli-Depth and full depth compared to previous works [113, 216].

- **Toom-Cook.**
  The Toom-Cook algorithm is a faster generalization of Karatsuba's method, and several existing quantum implementations are based on this approach.
  The first quantum Toom-Cook multiplier was proposed by Dutta et al. [47] in 2018. This initial design is based on Toom-Cook 2.5. By employing reversible pebble games to uncompute the intermediate results, this design was further improved in terms of Toffoli-Depth, Toffoli-Count, and qubit count.
  In 2021, Larasati et al. [122] investigated the quantum circuit for Toom-3 multiplication, which is expected to provide an asymptotically lower depth than the quantum Toom-2.5 design [47]. Their design achieves lower asymptotic complexity in terms of Toffoli-Depth and qubit count compared to Toom-2.5. However, it involves a larger number of Toffoli gates, primarily due to the implementation of the division operation.
  In 2023, Putranto et al. [168] proposed a higher-degree multiplier, the Toom-Cook 8-way multiplier, which has the lowest asymptotic performance and implementation cost. To design the Toom-Cook 8-way multiplier, they employed step-by-step computations, including splitting, evaluation, point-wise multiplication, interpolation, and recomposition, along with several strategies to reduce space and time requirements. As shown in Table 3.22, the Toom-Cook 8-way multiplier demonstrates space- and time-efficient multiplication. Moreover, Fig. 3.61 visualizes the performance of different quantum Toom-Cook multipliers.
- **Wallace Tree.**
  A multiplier circuit can also be achieved by using the Wallace tree multiplication technique [221], known as one of the fastest multiplier architectures. This efficient method cascades a sequence of full adders and half adders to sum the partial products. In 2021, Gayathri et al. [61] proposed the first quantum multiplier based on the Wallace tree multiplication architecture. Compared to the quantum conversion of existing reversible Wallace tree multiplier designs, this work implements a quantum integer multiplication circuit using an efficient quantum full adder, resulting in significant Toffoli-Count and Toffoli-Depth savings.

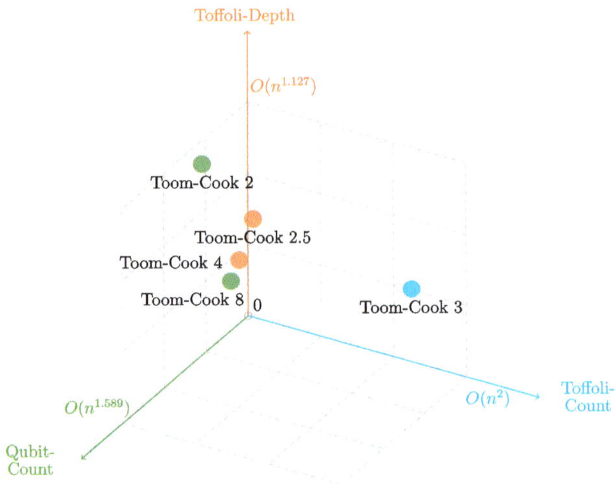

Fig. 3.61 Visualization of the performance of different Toom-Cook quantum multipliers

In 2023, Orts et al. [154] designed and analyzed a quantum circuit using the Wallace tree technique to compute the product of two integers. Their work involved studying existing circuits in the literature and optimizing the implementation of low-level operations using various techniques. Unlike the quantum Wallace tree multiplier by Gayathri et al. [61], which is limited to specific number sizes, this design can handle numbers of any size N and achieves better Toffoli-Count, Toffoli-Depth, and qubit count.

- **Schönhage-Strassen.**
  In classical computing, the Schönhage-Strassen algorithm is faster than the Karatsuba and Toom-Cook algorithms, sparking interest in its potential for quantum applications. The first Schönhage-Strassen-based multiplier was designed by Zalka in 1998 [245], leveraging the 2nd level FFT to reduce the computational cost for integer multiplication.
  In 2023, Nie et al. [149] designed a family of quantum circuits for integer multiplication based on the classical Schönhage-Strassen algorithm. Compared to the quantum Toom-Cook multiplier [47] and the quantum Karatsuba multiplier [64], this design offers a reduced gate count and circuit depth but requires more qubits.
- **Windowed.**
  Windowed arithmetic, unlike other multiplication methods, aims to reduce the number of operations by using lookup tables to merge operations. In 2019, Gidney [65] proposed windowed arithmetic circuits based on lookup tables. Resource estimation conducted using the Azure quantum resource estimator [85] indicates that the windowed multiplication design offers significant advantages in algorithm runtime and Toffoli requirements.

Table 3.23 An overview of prominent QFT-based quantum multipliers

Year	Article	Structure	Key innovations
2002	Beauregard [13]	Schoolbook	Propose the first schoolbook multiplier
2004	Nguyen [145]	Shift-and-Add	Propose the first shift-and-add multiplier
2012	Pavlidis et al. [165]	Shift-and-Add	Improve design with quantum-classical adder
2013	Maynard et al. [139]	Shift-and-Add	Propose multiply-add operation
2017	Ruiz-Perez et al. [177]	Shift-and-Add	Calculate controlled weighted sum
2018	Rines [175]	Shift-and-Add	Propose modular multiplier
2020	Zhang [247]	Problem-aware designs	Optimize depth of multiplier
2023	Zhan [246]	Problem-aware designs	Optimize multiplier with exponent encoding
2023	Ramezani et al. [170]	Convolution theorem	Reduce complexity with convolution theorem
2023	Kahanamoku-Meyer et al. [109]	Toom-Cook	Propose the first Toom-Cook multiplier

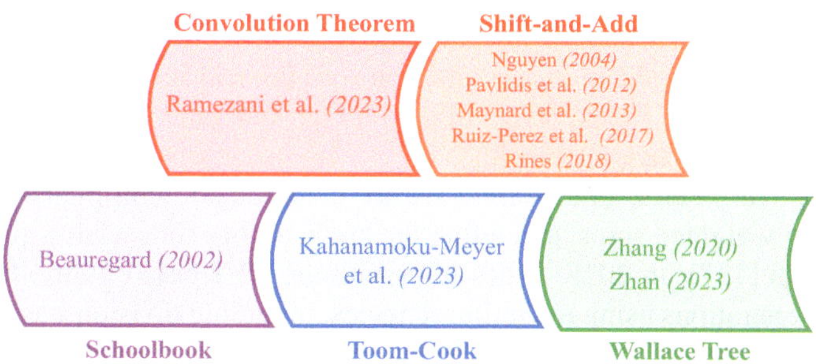

Fig. 3.62 General classification of QFT-based quantum multipliers articles as per structure and sorted by publication year

QFT-Based Circuit Designs

Quantum multiplications using QFT-based blocks have been discussed in the literature. Similar to the methods used in Clifford+T-based design, we categorize QFT-based multipliers into the following types, with lines of research listed in Table 3.23. A general classification diagram is provided in Fig. 3.62 to group each article based on its structure and sorted by publication year for QFT-based quantum multipliers.

- **Schoolbook.**
  Similar to Clifford+T-based multipliers, one of the most straightforward implementations of a multiplier is to construct it using sequences of adders. In 2002,

Beauregard introduced the first QFT-based multiplier for integer arithmetic with layers of quantum-classical QFT-based adders [13]. While the design is simple, it is inefficient and requires a large amount of resources. Additionally, Beauregard's schoolbook method works only for integer multiplication between a quantum factor and a classical constant.

- **Shift-and-Add.**
  Another typical design for a QFT-based multiplier is based on the Shift-and-Add principle. Shift-and-add steps naturally correspond to the binary representation of auxiliary controlled qubits and addition arithmetic. Multiplication of two factors involves setting one factor as the control qubit, which performs additions to the other factor in the Fourier basis. Therefore, multipliers are constructed with controlled QFT-based adders and further decomposed into a QFT block, controlled-controlled phase gates, and an inverse QFT block.
  The first quantum multiplier based on the shift-and-add principle and Draper's quantum-quantum adder was proposed by Nguyen in 2004 [145]. Pavlidis and Gizopoulos improved it in 2012 by replacing Draper's quantum-quantum adder with Beauregard's quantum-classical adder for multiplying a classical constant to the quantum state [165]. In this chapter, the structure of the Fourier-based controlled quantum-classical adder was used as a subroutine for constant multiplications.
  Another novel Fourier-based gate, known as the multiply-add operation, was proposed by Christopher Maynard and Einar Pius in the same year [139]. They merged multiplication and addition for classical integers into a single unit consisting of a QFT block, layers of parallel fan-out gates or single-qubit rotation gates, and an inverse QFT block.
  In 2017, Ruiz-Perez and Garcia-Escart constructed multipliers to calculate controlled weighted sums, extending the applicability for scalable quantum multiplications [177]. In 2018, Rines improved the QFT-based multiplier to handle modular operations using reduction schemes, including division-based reduction, adaptations of classical Montgomery multiplication, and Barrett reduction [175].
- **Convolution Theorem.**
  The convolution theorem states that the convolution of two discrete vectors in the time domain is equivalent to their pairwise multiplication in the frequency domain. This equivalence is quantified by the Fourier transformation of the convolution of two functions and the product of their Fourier transforms.
  Inspired by this, Ramezani et al. proposed an efficient quantum algorithm for integer multiplication in 2023. By leveraging quantum amplitude amplification, they reduced the time complexity to $O(\sqrt{n}\log^2 n)$, outperforming the best-known classical algorithm with time complexity $O(n\log n)$ [170]. Since the convolution theorem has no counterpart in Clifford+T-based multipliers, this design is considered a novelty enabled by the power of the Fourier transform. Nevertheless, exploring the application of various theorems related to Fourier transformation that link complex arithmetic to attainable operations remains an open and interesting problem.

- **Toom-Cook.**
  Kahanamoku-Meyer et al. proposed a fast quantum integer multiplication algorithm with zero ancillas based on the classical Toom-Cook algorithm [109]. They developed an efficient phase rotation algorithm using Draper's QFT-based adder to perform phase product operations and integrated it into the classical Toom-Cook algorithm, achieving sub-quadratic time complexity for multiplications.
- **Problem-Aware Designs.**
  Some multiplier designs leverage problem specifications to optimize the circuit extensively. We briefly summarize several intriguingly developed circuits with improved efficiency and reduced applicability, as these designs are problem-aware or based on specific assumptions.
  For example, in 2023, Junpeng Zhan proposed the quantum multiplier based on exponent adder (QMbead) as a novel approach for quantum multipliers [246]. This approach consists only of an in-place or out-of-place QFT-based adder. Multiplication results are obtained through measurement outcomes. Although this method is efficient and cost-effective in arithmetic operations, it has a strong restriction on the data encoding procedure. The initial states must be the superposition of all exponents of the two input factors, which is highly expensive in practice [89, 131].
  Another problem-aware design for QFT-based multipliers involves starting from a conventional quantum multiplier and applying compilation methods to optimize gate depth. An example of this can be found in [247].

3.6.5 Quantum Division

In classical computing, division algorithms can be broadly classified into two categories based on their computational efficiency—fast and slow methods. Fast division algorithms are designed to minimize the number of computational steps required for obtaining the quotient and remainder. Two notable examples of such algorithms are the Newton-Raphson and Goldschmidt [69] which use iterative methods to rapidly converge on a result. Conversely, slow division algorithms, such as long division, restoring division, non-restoring division [191], and SRT division [87], are typically inefficient but offer a straightforward approach to understanding and implementation. Quantum division builds upon the foundational work of classical division, as detailed in the following sections.

Clifford+T-based Designs
Table 3.24 provides a categorization of quantum division using Clifford+T-based quantum dividers, including information on their publication years, articles, structures, and key innovations. To better illustrate each quantum divider's efficiency (fast or slow division) and its implementation type (Clifford+T or QFT), a quadrant analysis chart is provided in Fig. 3.63.

Table 3.24 An overview of prominent Clifford+T-based quantum dividers

Year	Article	Structure	Key innovations
2019	Thapliyal et al. [210]	Restoring	Propose the first restoring divider
2019	Thapliyal et al. [210]	Non-restoring	Propose the first non-restoring divider
2021	Gayathri et al. [61]	Goldschmidt	Propose the first Goldschmidt divider
2022	Gayathri et al. [60]	Newton-Raphson	Propose the first Newton-Raphson divider
2022	Yuan et al. [242]	Long division	Reduce overall cost of designs [210]
2024	Wang et al. [230]	Non-restoring	Apply comprehensive DSE to reduce cost
2024	Sajadimanesh et al. [181]	Restoring	Implement on quantum computers
2024	Sajadimanesh et al. [181]	Non-restoring	Implement on quantum computers
2024	Orts et al. [155]	Long division	Reduce Toffoli-Count and Toffoli-Depth

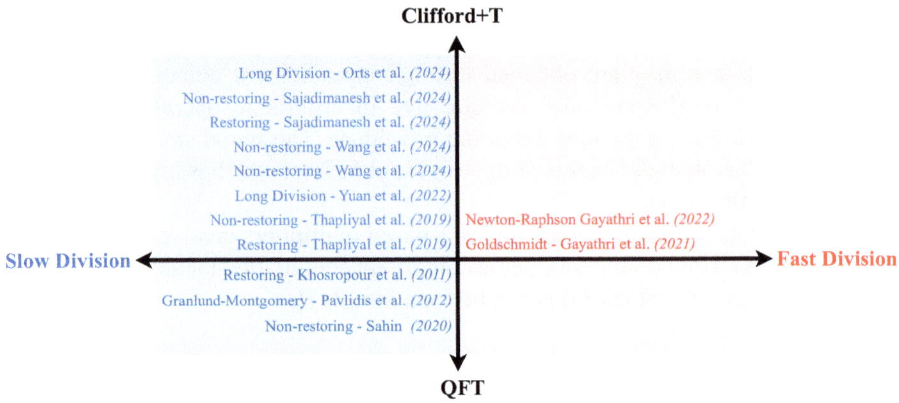

Fig. 3.63 Quadrant analysis chart for quantum dividers based on efficiency and implementation type

- **Slow Division.**
  In 2019, Thapliyal et al. [210] introduced quantum restoring and non-restoring dividers, inspired by classical slow division algorithms. This marked a significant advancement in the quantum division domain, presenting initial designs based on the Clifford+T set. However, the T-Depth calculation in this work uses a nonstandard method, so we disregard the reported cost. The corresponding correction can be found in another work [230].
  Quantum long division remained unexplored until 2022 when Yuan et al. [242] introduced a novel fault-tolerant quantum divider based on the long division algorithm using Clifford+T gates, leveraging quantum comparators and subtractors. This design exhibits significant reductions in Toffoli-Count, Toffoli-Depth, and Qubit count compared to existing quantum restoring and non-restoring dividers [210].
  Drawing from previous quantum slow division designs, Wang et al. [230] concentrate on improving the performance of quantum slow dividers by optimizing the design of their subblocks. By thoroughly exploring the design space of

state-of-the-art quantum addition building blocks, this work achieves notable advancements over previous quantum slow division designs. Moreover, this work underscores the importance of adopting a systematic approach to design space exploration (DSE).

In the same year, Sajadimanesh et al. [181] used dynamic circuits and approximate computing to implement quantum restoring and non-restoring division circuits on quantum computers. Dynamic circuits were used for mid-circuit measurement, reducing the number of required qubits and enhancing fidelity. Additionally, approximate computing was employed to mitigate noise inherent in quantum hardware. The division circuits were executed on the IBM quantum NISQ computer, demonstrating its capability to overcome quantum noise and yield meaningful results.

In 2024, Orts et al. [155] introduced a novel divider circuit based on long division, aimed at reducing the number of T gates while maintaining a comparable qubit count. This approach involves introducing variant minor T-optimized circuits. As a result, the Toffoli-Depth and Toffoli-Count of this design surpass all previous quantum slow division-based dividers [210, 242].

- **Fast Division.**
  In 2021, Gayathri et al. [61] introduced the first quantum fast division circuit based on the Goldschmidt division algorithm, using the IEEE 754 single-precision format. However, the overall cost of quantum fast division remains significantly higher compared to slow division designs. One year later, Gayathri et al. [60] proposed a quantum floating-point divider based on the Newton-Raphson method. Table 3.25 shows that their division circuit yields substantial savings in T-Count and qubit count compared to state-of-the-art fast division algorithms such as the quantum Goldschmidt divider [61] and the quantum conversion of an efficient reversible floating-point divider based on the Goldschmidt algorithm [7]. This advantage is brought by the improved structure design and more efficient Toffoli decomposition method into T gates.

QFT-Based Circuit Designs

Quantum dividers based on Fourier transformation have characteristics similar to those based on Clifford+T gates. Here we briefly list three exemplars based on slow division algorithms in Table 3.26. The first QFT-based divider was proposed by Khosropour in 2011 using a restoring division algorithm with QFT-based adders and subtractors [114]. In 2012, Pavlidis used Draper's quantum-quantum adder and Beauregard's quantum-classical structure to develop a quantum divider based

Table 3.25 Performance analysis of prominent fast quantum division designs based on IEEE 754 single-precision format

Article	Structure	Qubit count	T-Count	T-Depth
Gayathri et al. [61]	Goldschmidt	29,074	22,7920	59,916
AnanthaLakshmi et al. [7]	Goldschmidt	30,008	117,187	17,850
Gayathri et al. [60]	Newton-Raphson	23,996	93,376	13,506

Table 3.26 An overview of prominent QFT-based quantum dividers

Year	Article	Structure	Key innovations
2011	Khosropour et al. [114]	Restoring	Propose QFT-based restoring divider
2012	Pavlidis et al. [165]	Granlund-Montgomery	Realize division by a constant
2020	Sahin [254]	Non-restoring	Propose QFT-based non-restoring divider

on the Granlund-Montgomery classical division by constant algorithm [165]. In 2020, Sahin proposed a non-restoring QFT-based division circuit composed of basic blocks including QFT-based absolute value, QFT-based modular subtractor, two's complement, and modular adder [254].

3.6.6 *Quantum Modular Exponentiation*

Quantum modular arithmetic includes all basic arithmetic operations with bounded variables in a periodic manner. All four basic arithmetic operations—addition, subtraction, multiplication, and division—can be performed in a modular structure. Another infrastructural operation is the modular exponentiation. Although exponentiation itself is a transcendental function that may be complex to implement, quantum algorithms such as Shor's factorization rely heavily on the realization of modular exponentiation. In most architectures, basic quantum arithmetic serves as a fundamental building block for constructing modular exponentiation. Considering the above, this section focuses only on quantum modular exponentiation designs composed of basic modular arithmetic. Unlike the taxonomic approach above, here we put Clifford+T designs as well as QFT designs in Table 3.27. Interestingly, we also investigate several works that include both Clifford+T and QFT designs, as listed in Table 3.27 under Both. For clarity, a Venn diagram for modular exponentiation is provided in Fig. 3.64.

- **Clifford+T-Based Designs.**
  In 1996, Clifford+T-based quantum modular exponentiation designs emerged. These designs rely on quantum multiplication circuits built from ripple-carry adders, which, while simple, are the slowest method. As the first work on this topic, Vedral et al. [218] made significant contributions by explicitly constructing quantum networks to perform elementary arithmetic operations, including modular exponentiation. They constructed full modular exponentiation from smaller quantum addition building blocks, with most of the time spent in $20n^2 - 5n$ ripple-carry adders. In this design, the auxiliary memory required grows linearly with the size of the number to be factorized.
  In the same year, Beckman et al. [14] developed a similar full modular exponentiation architecture based on ripple-carry adders. Their work differed from Vedral

Table 3.27 An overview of prominent quantum modular exponentiation designs

Year	Article	Basis	Key innovations
1996	Vedral et al. [218]	Clifford+T	Use arithmetic for modular exponentiation
1996	Beckman et al. [14]	Clifford+T	Use arithmetic for modular exponentiation
2000	Cleve et al. [35]	QFT	Implement QFT in parallel
2002	Beauregard [13]	QFT	Realize modular exponentiation with adder
2004	Fowler et al. [54]	QFT	Consider linear nearest neighbor constraint
2005	Van Meter et al. [217]	Clifford+T	Boost the efficiency with concurrency
2010	Fu et al. [57]	QFT	Utilize semiclassical QFT
2012	Markov et al. [136]	Clifford+T	Use predefined parameters for specific cases
2012	Pavlidis et al. [165]	QFT	Implement using modular multipliers and divider
2013	Smolin et al. [196]	QFT	Oversimplify Shor's algorithm in experiments
2014	Lin et al. [127]	Both	Propose benchmarks for both bases
2018	Young et al. [241]	QFT	Study comb function using Hadamard transform
2018	Häner et al. [84]	Clifford+T	Apply piecewise polynomial approximation
2019	Das et al. [41]	Clifford+T	Map logic designs to quantum architectures
2019	Gidney [65]	Clifford+T	Introduce lookup tables
2021	Liu et al. [130]	Clifford+T	Minimize the number of CNOT gate
2022	He et al. [91]	Both	Proposes a BER protocol
2023	Yuan et al. [243]	QFT	Design comparator-based modular exponentiation
2023	Singleton [194]	QFT	Use predefined parameters for specific cases
2023	Atchade-Adelomou et al. [9]	QFT	Summarize in-place and out-of-place structures

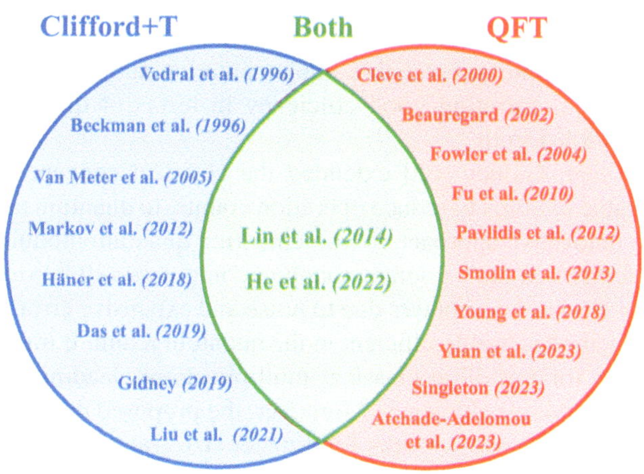

Fig. 3.64 Venn diagram for quantum modular exponentiation designs

et al.'s by more aggressively leveraging classical computation and presenting several optimizations and trade-offs between space and time.

In 2005, Van Meter et al. [217] introduced logical concurrency into the existing quantum modular exponentiation framework. By exploiting concurrency, they aimed to reduce the execution time of modular exponentiation, thereby enhancing Shor's algorithm. Their work provided a detailed analysis of the impact of architectural features and concurrent gate execution on quantum modular exponentiation. The main contributions focused on paralleling arithmetic execution through improved adders, concurrent gate execution, and an optimized overall algorithmic structure.

Apart from the general modular exponentiation structures capable of handling all possible inputs, one can also create specialized quantum circuits for modular exponentiation with predefined inputs to reduce overall costs. Following this idea, Markov et al. [136] proposed quantum circuits for modular exponentiation in specific input cases in 2012, which provided constant-factor improvements and a significant constant additive term improvement for few-qubit circuits.

In 2018, Häner et al. implemented modular exponentiation along with other transcendental arithmetic functions such as inverse of *square root*, *sin*, and *arcsin* using piecewise polynomial approximations [84]. They proposed an efficient circuit composed of adder and multiplier to evaluate many polynomials in parallel. The authors referred to adders and multipliers using Clifford+T-based designs, although the blocks are independent of elementary implementations. Therefore, we categorize this as a Clifford+T-based modular exponentiation design.

Additionally, mapping practical logic designs to quantum architectures for implementing quantum modular exponentiation is a viable approach. In 2019, Das et al. [41] used synthesis tools to generate a quantum circuit from a Verilog implementation of exponentiation functions. They converted the logic circuit to a corresponding reversible implementation and mapped it to quantum architectures, demonstrating good efficiency in terms of qubit count, Toffoli-Count, and Toffoli-Depth.

In the same year, Gidney [65] extended the classical concept of windowing, which uses table lookups to reduce operation counts, to quantum arithmetic. This provided an innovative approach to implementing quantum modular exponentiation. Considering that fault-tolerant quantum computers will be significantly less efficient than classical computers due to noise and expensive error correction, the windowing method is highly efficient in the quantum realm. It trades 20 quantum multiplications for a million classical multiplications, leading to a significant improvement in overall efficiency. Moreover, the proposed design exhibits lower Toffoli-Count than previous works for register sizes ranging from tens of qubits to thousands of qubits.

In contrast to previous efforts on minimizing Toffoli-Depth, Toffoli-Count, or qubit count, Liu et al. [130] aimed to reduce the number of CNOT gates, which significantly impacts the running time on an ion trap quantum computer. Their work achieves basic arithmetic implementations with the lowest known number of CNOT gates. By accumulating intermediate data and employing the windowing technique, they also achieve a low CNOT-count construction for a quantum modular exponentiation design.

- **QFT-Based Designs.**
  Modular exponentiation circuits using QFT-based blocks have been well discussed in the literature as a key part of Shor's factorization algorithm. Here, we briefly review some of the mainstream works using QFT for modular exponentiation.
  The most typical model was proposed by Beauregard in 2002 [13]. It includes constant number arithmetic operations such as controlled modular adders, controlled modular multipliers, SWAP gates, a QFT block, and an inverse QFT block. In reference [54], Fowler et al. discussed implementing modular exponentiation with the constraint that only linear nearest-neighbor interactions are available. They modified Shor's algorithm by interpolating sequences of SWAP gates and controlled SWAP gates, making it more practical for real hardware settings. Park and Ahn further improved this method for the QFT circuit with linear nearest-neighbor architecture by reducing the CNOT count [161]. In 2012, Pavlidis and Gizopoulos proposed the fast quantum modular exponentiation architecture combining quantum-classical modular multipliers and a quantum divider by a constant number [165]. In 2023, Atchade-Adelomou and Gonzalez implemented quantum modular exponentiation with out-of-place operations using PennyLane software, considering intermediate-scale quantum executions [9]. In Singleton's work [194], the factorization problem is mapped onto another mathematical problem in number theory that finds the period of modular exponential functions. Using the continued fractions algorithm, the complexity of Shor's factorization problem is reduced by directly extracting solutions from approximately measured phase angles, improving the design for the modular exponentiation circuit.
  An alternative framework for implementing modular arithmetic is based on quantum comparators. A comparator is an oracle that takes two factors, a and b, as input and determines whether a is larger than b or vice versa. Modular arithmetic with respect to a and b is then realized by a comparator followed by some basic blocks. For example, in 2023, Yuan et al. designed a novel QFT-based comparator using Beauregard's quantum-classical adder to realize modular addition arithmetic. They then constructed quantum modular multipliers and modular exponentiation circuits [243]. Although Yuan et al.'s design requires more QFT blocks compared to the original Beauregard circuit, it reduces the number of auxiliary qubits and improves execution on small-size quantum computers.
  Additionally, there are several advancements using optimization techniques to reduce the resource cost for QFT and improve modular exponentiation algorithms in practice. For example, in 2000, Cleve and Watrous designed a protocol for fast parallel implementation of QFT by copying Fourier states and estimating phases in parallel, reducing the complexity to logarithmic depth [35]. In 2018, Young et al. simplified Shor's algorithm for a specific transformation of the

comb function and replaced QFT with quantum Hadamard transformation, which requires only controlled rotation gates with angles of 0 and π, thus lessening the experimental burden [241].

Another type of optimized modular exponentiation circuits is obtained using semiclassical QFT, proposed by Griffiths and Niu [73]. This structure transforms controlled quantum operations followed by measurements into classical controlled operations and postprocessing steps, reducing the number of qubits. Inspired by this, Fu et al. utilized semiclassical QFT to speed up the implementation taking into account the non-adjacent form representation [57]. In 2013, Smolin et al. compiled Shor's factorization algorithm for experimental demonstration [196]. Using semiclassical QFT and qubit reuse, they showed both theoretically and experimentally that Shor's algorithm can be realized by small circuits for any composite number $N = pq$. In this complied Shor's algorithm, modular exponentiation is trivial since the exponent only denotes a single qubit and the base is 2. It is suitable for hardware platforms since it corresponds to the simplest QFT-based adder, made of a Hadamard gate as a single-qubit QFT block, a controlled-NOT gate as the controlled phase gate, and another Hadamard gate as a single-qubit inverse QFT block.

- **Adaptable Designs.**
  In 2014, Lin et al. [127] proposed Qlib, a quantum module library designed to serve as a suite of benchmarks for quantum logic. The modular exponentiation circuit within Qlib is constructed using a hierarchical decomposition based on a full adder. Interestingly, this work includes both Clifford+T-based and QFT-based quantum modular exponentiation designs, utilizing Cuccaro's Clifford+T-based adder [40] and Draper's QFT-based adder [45], respectively. In 2022, He et al. designed a quantum modular exponentiation circuit using the binary-exponent-based recombination (BER) protocol [91]. They proposed three implementations of the BER protocol with modular multipliers and modular adders, ranging from Clifford+T-based designs to QFT-based designs.

3.6.7 *Future Directions*

In the future, several areas hold promise for advancing quantum arithmetic:

- **Windowed Arithmetic:** Most current work focuses on improving arithmetic design circuits, but enhancing lookup table designs also has the potential to advance quantum arithmetic, especially with the rapid development and widespread use of quantum windowed arithmetic [65]. Future research could explore the optimization of both quantum arithmetic blocks and lookup tables to achieve better performance.
- **Constant Arithmetic:** Compared to variable arithmetic, arithmetic operations involving a variable and a constant, such as reciprocal and squaring, are less explored. While several existing designs, such as those by Markov et al. [136]

and Singleton [194], have made contributions, the potential efficiency gains from replacing a variable with a constant remain largely unexplored and represent a valuable area for future research.
- **Arithmetic in Ternary and Other Bases:** In classical arithmetic, multivalued logic, especially ternary logic, has been extensively explored. Ternary representation in quantum computing could lead to smaller circuit designs with fewer inputs and outputs, reducing the number of qubits needed and enhancing overall efficiency. While most quantum arithmetic designs focus on binary designs, several works have concentrated on ternary reversible designs [8, 51, 159]. There is significant potential in exploring arithmetic in other bases, such as ternary logic.

Moreover, quantum arithmetic circuits are expected to play a crucial role in several emerging applications. The growing demand for quantum arithmetic as foundational building blocks is evident in various areas, including but not limited to the following applications:

- **Quantum Computation of Classical Functions:** From Shor's algorithm to activation functions, arithmetic designs have performed well in speeding up classical computational tasks. With the advancement of quantum computers, quantum arithmetic circuits have potential advantages in computing functions that are tough for classical computers.
- **Quantum Oracle Realizations:** Fault-tolerant quantum algorithms often reply on linear or nonlinear algebra with the block-encoded oracles. Given that the explicit constructions are often built from basic operations, quantum arithmetic circuits are hoping to be of vital for accessing classical data and upload them to the quantum routines.

3.6.8 Section Summary

In conclusion, the field of quantum computing has significantly evolved in recent decades, yielding remarkable advancements. Quantum arithmetic circuits, fundamental components for constructing quantum algorithms, have garnered considerable attention. While numerous studies on efficient quantum arithmetic circuit designs exist, there has been a pressing need for a systematic review to summarize these achievements. In this chapter, we conducted a comprehensive investigation into quantum arithmetic designs, covering basic operations such as addition, subtraction, multiplication, division, and modular exponentiation. Using various evaluation metrics, we assessed the efficiency of quantum arithmetic circuits based on Clifford+T set and QFT. We also proposed avenues for future research. Our work provides researchers and engineers with a comprehensive reference to better understand and apply the latest advancements in quantum arithmetic designs, fostering further progress in quantum computing.

3.7 Chapter Summary

This chapter presents our research on efficient quantum arithmetic circuit designs, emphasizing their impact on quantum efficiency, which is essential for enhancing the scalability of quantum computing.

The first work introduces a higher radix layer that significantly reduces T-Depth and qubit usage, surpassing existing designs by requiring $O(n) - O(\frac{n}{r})$ qubits and achieving optimal T-depth of $O(r) + O(\log \frac{n}{r})$. In the second work, we developed a parallel prefix adder using Ling expansion, achieving a T-Depth of $O(\log \frac{n}{2})$. This improvement addresses large fan-out challenges, optimizing the overall efficiency of quantum addition. In the third work, our analysis of 160 carry-propagation structures led to a design with a Toffoli-Depth of $\log n + O(1)$, reducing depth by 50% compared to previous designs, validated through theoretical and simulation studies. Research 4 proposes a specific framework converts multiple 2-addend operations within modular addition into 3-addend ones, yielding a Toffoli-Depth reduction of up to 33.33%, while maintaining qubit count, showcasing its potential for efficient modular arithmetic. In work 5, we achieved a Toffoli-Depth decrease of up to 93.90% and significant reductions in Toffoli (up to 92.12%) and qubit count (up to 99.38%), highlighting the effectiveness of targeted optimizations for quantum dividers. The chapter concludes with a structured overview of recent developments in quantum addition, subtraction, multiplication, division, and modular exponentiation, suggesting future research directions for improving circuit efficiency.

Key Takeaways

1. **Quantum Higher Radix Adder:** Utilized a higher radix layer to reduce T-Depth to $O(r) + O(log\frac{n}{r})$ and qubit usage to $O(n) - O(\frac{n}{r})$.
2. **Quantum Ling Adder:** Achieved T-Depth of $O(\log \frac{n}{2})$ using Ling expansion, effectively addressing large fan-out challenges.
3. **Optimal Depth Adder:** Analyzed 160 adder designs and developed an optimal depth adder with Toffoli-Depth $\log n + O(1)$, reducing depth by 50%.
4. **Modular Addition Framework:** Converted 2-addend operations to 3-addend ones, reducing Toffoli-Depth by up to 33.33% while maintaining qubit count.
5. **Quantum Divider Optimization:** Achieved up to 93.90% Toffoli-Depth reduction, 92.12% fewer Toffoli gates, and 99.38% fewer qubits through design space exploration.
6. **Overview:** Provided an overview of recent advancements in quantum arithmetic and suggested future research directions for improving quantum circuit efficiency.
7. These designs enhance quantum arithmetic efficiency, establishing a foundation for scalable quantum computing.

Chapter 4
Applications of Quantum Arithmetic Designs

There is no such thing as a special category of science called applied science; there is science and its applications, which are related to one another as the fruit is related to the tree that has borne it.

–Louis Pasteur

This chapter demonstrates the impact of efficient quantum arithmetic circuits on two important application domains: quantum machine learning (QML) and quantum cryptanalysis. Firstly, we discuss how to map our design onto real quantum hardware architectures, specifically focusing on the verification of our design using NISQ devices, which are currently the only available hardware architecture in quantum computing. Afterward, each application highlights the potential of quantum arithmetic to enhance overall computational efficiency across diverse domains. The following sections provide the details of our research.[1]

4.1 Mapping to NISQ Devices

4.1.1 Step 1: Programming

The first step in mapping our quantum designs to NISQ hardware involves programming quantum circuits using existing quantum languages, such as Qiskit. These languages offer essential tools for representing quantum implementations, facilitating a seamless transition from theoretical designs to practical applications. This basic step enables us to represent quantum circuits effectively by employing qubits and quantum gates defined within quantum programming frameworks.

[1] This chapter includes our papers [227, 252].

S. Wang, A. Chattopadhyay, *Quantum Arithmetic*, Computer Architecture and Design Methodologies, https://doi.org/10.1007/978-981-95-6039-4_4

4.1.2 Step 2: Simulating

The second step involves testing the programmed quantum designs using quantum simulators. Quantum simulators are essential for conserving valuable quantum resources, as they simulate the behavior of quantum hardware in a noise-free or customizable noise environment, which allows for thorough verification of the design's correctness before it is applied to real-world quantum hardware. By using quantum simulators, potential issues can be identified and resolved, ensuring that the quantum circuits are well-prepared for execution on actual quantum hardware.

4.1.3 Step 3: Mapping

The final step involves mapping the validated designs to real quantum hardware. This stage includes deploying the circuits to cloud-based quantum platforms, such as the IBM Quantum Platform, which allows execution on physical quantum processors. For most designs in this book, the *ibm_brisbane* quantum device was used, with its qubit layout shown in Fig. 4.1.

This mapping process is vital for circuit performance, as it involves selecting appropriate qubits and determining optimal routing paths to minimize quantum gate errors and maintain coherence. The IBM platform supports automatic optimizations through built-in functions, which help streamline the mapping and execution process, ensuring efficient deployment on the hardware.

In this book, we successfully mapped all our designs to quantum hardware. Each arithmetic design was tested with random inputs and verified to function correctly. It is noteworthy that our research contributed to the development of Qrisp,[2] which is a high-level programming language designed to facilitate the creation and compilation of quantum algorithms. Specifically, Qrisp offers user-friendly tools and intuitive abstractions. Its features include: (1) Intuitive quantum program design, (2) High-level quantum programming, (3) Efficient quantum algorithm implementation, and (4) Extensive documentation and examples.

4.2 Applications in Quantum Machine Learning

In our exploration of quantum machine learning, we conducted a design [252] titled "Efficient quantum circuits for machine learning activation functions including constant T-depth ReLU." In this chapter, I proposed an efficient quantum design of

[2] https://qrisp.eu

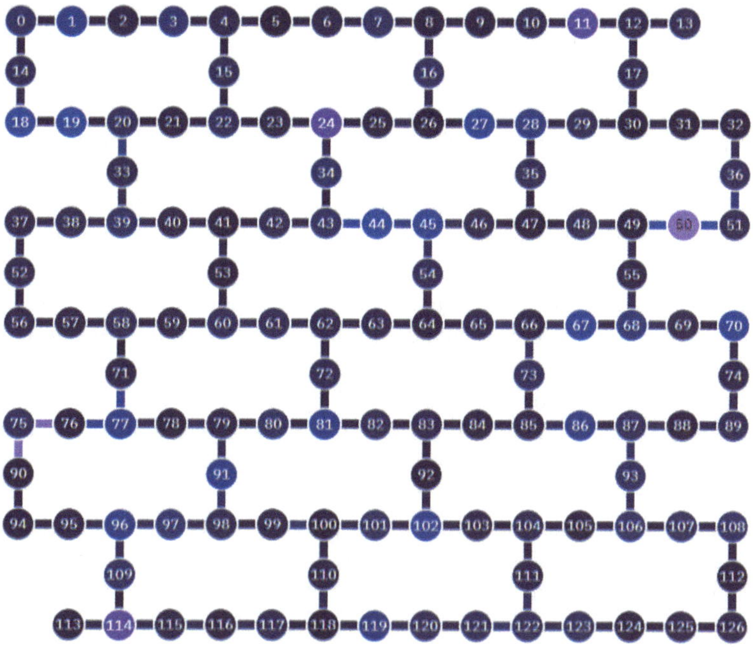

Fig. 4.1 Qubits layout of the *ibm_brisbane* quantum computer

the Leaky ReLU activation function, which serves as a quantum machine learning application involving arithmetic operations.

In recent years, quantum machine learning (QML) has increasingly captured the interest of researchers. Among the components in this domain, activation functions hold a fundamental and indispensable role. This research focuses specifically on the development of quantum circuits for the Leaky ReLU activation function, aimed at integration into fault-tolerant quantum computing architectures with an emphasis on minimizing T-depth. Specifically, I present a novel implementation of leaky ReLU activation function, achieving a constant T-depth of 8. This study represents a significant advancement toward enhancing the practicality and application of quantum machine learning.

4.2.1 Introduction

Currently, machine learning (ML) is attracting a lot of attention, driven by its widespread adoption across various industries. Recent breakthroughs in natural language processing models, exemplified by GPT4 [153], as well as advancements in image classification and generation [3, 125, 148, 240], highlight the rapid advancements of this field. An important building block of machine learning architectures is feed-forward neural networks [78], which are collections of layers of activation functions and linear transformations. In previous research, many different activation functions have been proposed, such as Sigmoid, Softmax, and ReLU, which are successfully utilized in ML applications.

Quantum machine learning [186] has shown progress in both adapting common machine learning algorithms to run on future quantum computers and using machine learning techniques to learn about quantum systems [166]. Some examples include quantum neural networks (QNN) [1], quantum support vector machines (QSVM) [174], quantum principal component analysis (QPCA) [133], variational quantum eigensolver (VQE) [212], parameterized quantum circuits (PQC) [15], and variations of the quantum approximate optimization algorithm (QAOA) [251].

Recently, numerous research findings on QNNs based on quantum circuit implementations have been proposed [36, 185, 203, 222, 249]. It turns out that the implementation of activation functions, using quantum circuits, is fundamental and crucial in QNNs. Various methods have been utilized to implement activation functions, including quantum phase estimation [163, 187], Taylor series expansion [137], quantum oracle synthesis [238], and polynomial networks [132]. However, to the best of the authors' knowledge, there are no dedicated circuit designs for implementing activation functions with a focus on fault tolerance so far. In this chapter, we address this gap by constructing a quantum circuit using Clifford+T gates. We specifically focus on minimizing the T-depth of the circuits, considering the high cost associated with fault-tolerant implementations of the T gate and the limitation imposed by the coherence time of the quantum device [71]. In particular, for some activation functions that have a high implementation complexity, such as the ReLU [190], approximating them using polynomials would result in high polynomial degrees [67]. We have specifically designed quantum circuits to implement them, significantly reducing the T-depth.

In particular, the first author Wei Zi proposes a quantum circuit to implement the ReLU function with a constant T-depth of 4, without using ancillary qubits. If the qubit connectivity is constrained to a 2D grid, the T-depth of the circuit for the ReLU function remains unchanged. Additionally, I implement Leaky ReLU functions using a quantum circuit with a constant T-depth of 8. The results of ReLU and Leaky ReLU are summarized in Table 4.1. Here, we have optimized both the T-depth and ancilla count to an optimal level. We attain a constant T-depth (independent of input size) simultaneously without the employment of any ancillary qubits. This implies that the ReLU and Leaky ReLU functions can be implemented at an extremely low cost in fault-tolerant quantum computation through our construction [71].

Table 4.1 The T-depth of the quantum circuit implementing ReLU in both scenarios: with and without the constraint of qubit connectivity limited to a 2D grid. The T-depth for Leaky ReLU is also presented

Activation function	ReLU	ReLU (on 2D grid)	Leaky ReLU
T-depth	4	4	8

We focus on the optimization of T-depth of the quantum circuit, that is, represents the running time of the quantum circuit in fault-tolerant quantum computation [71]. However, even though the physical implementation of quantum error correction has progressed greatly in recent years [4, 20, 119], it is still hard to use real devices to implement our circuit fault-tolerantly. Thus we only show the construction of the quantum circuit and analyze the performance. The Qiskit implementation of quantum circuits for activation functions discussed in this work is open-sourced on GitHub.[3]

4.2.2 *Activation Functions*

Activation functions are integral components in machine learning and deep learning, providing the crucial non-linearity required for effective learning. In the classical world, several activation functions are commonly used including ReLU, Leaky ReLU, Sigmoid, SoftMax, Tanh, Swish, exponential linear unit (ELU), and Gaussian linear unit (GELU) [93, 190]. Specifically, the formulas for ReLU, Leaky ReLU, and sigmoid are as follows:

- **ReLU:** $f(x) = \max(x, 0)$.
- **Leaky ReLU** ($0 < \alpha < 1$)**:**
  $f(x) = \max(x, \alpha x) = \begin{cases} x & x \geq 0 \\ \alpha x & x < 0 \end{cases}.$
- **Sigmoid:** $f(x) = 1/(1 + e^{-x})$.

For a quantum circuit U_f that can implement function f, it is defined as

$$U_f|x\rangle|0\rangle_o = |x\rangle|f(x)\rangle.$$

Here, $|x\rangle$ represents the quantum state that encodes the input of function f, $|0\rangle_o$ represents the output qubits used to store the result $f(x)$, and $|f(x)\rangle$ represents the quantum state that encodes the output $f(x)$. The main objective of this work is to construct U_f for various activation functions using Clifford gates and T gates. Specifically, we utilize the quantum gates from the Clifford+T gate set $\{CNOT, H, S, T\}$ to build the circuits. The commonly used X gate can be decomposed into four Clifford gates: $X = HSSH$. Due to the higher fault-tolerant implementation cost of the T gate compared to Clifford gates, our focus lies in minimizing the T-depth, which represents the circuit's runtime [71]. In our construction, three types of ancillary qubits may be utilized. Clean ancillary qubits are qubits initialized to $|0\rangle$ and reset to $|0\rangle$ after their use. Dirty ancillary qubits do not require the initial state but need to be reset after their use. Garbage qubits are initialized to $|0\rangle$ and do not need to be reset after use.

[3] https://github.com/ziwei1998/quantum-circuit-for-activation-functions

4.2.3 Methodology

Here, we discuss the details of the quantum implementation for the Leaky ReLU activation function.

The definition of the Leaky ReLU function is as follows:

$$f(x) = \begin{cases} x & x \geq 0 \\ \alpha x & x < 0,\ 0 < \alpha < 1 \end{cases}.$$

Definition 4.1 A circuit C_n can implement the n-bit Leaky ReLU function, which means that

$$C_n|\boldsymbol{x}\rangle|0\rangle_m = |\boldsymbol{x}\rangle|\max(\boldsymbol{x}, \alpha\boldsymbol{x})\rangle,$$

where $\boldsymbol{x} \in \{0, 1\}^n$ represents a fixed-point number, and the circuit C_n consists of Clifford+T gates. Here, α is a predetermined parameter chosen from the set $\{0.125, 0.0625, 0.03125, 0.015625\}$. The number of output qubits m is determined by the number of bits required to store $\max(\boldsymbol{x}, \alpha\boldsymbol{x})$.

Note that we choose α from the set $\{2^{-3}, 2^{-4}, 2^{-5}, 2^{-6}\}$. Therefore, $m = n+3, n+4, n+5$, or $n+6$ is sufficient to store the result $\max(x, \alpha x)$ exactly, depending on the specific choice of α. The input and output can be coded using either 2's complement or true form. The quantum circuit that implements the Leaky ReLU function will vary depending on the chosen coding method. For example, if x is coded as 1011, then $0.125x$ should be coded as 1000.011 and 1111.011 in true form and 2's complement, respectively. Note that the n and m qubits are needed to store the input and output, respectively. When the input has n bits, we will use $|0\rangle_1|0\rangle_2 \cdots |0\rangle_n$ to store the integer part and $|0\rangle_{n+1} \cdots |0\rangle_m$ to store the decimal part.

Theorem 4.1 *There exists a quantum circuit C_n capable of implementing the n-bit Leaky ReLU function with a constant T-depth of 8 without ancillary qubits. Additionally, the circuit depth and size are $O(\log n)$ and $O(n)$, respectively.*

To implement the Leaky ReLU function, we output x if the sign bit is 0, and we output αx if the sign bit is 1. It is worth noting that we choose α from the set $\{0.125, 0.0625, 0.03125, 0.015625\}=\{2^{-3}, 2^{-4}, 2^{-5}, 2^{-6}\}$. Therefore, αx can be easily obtained from x by performing a bit shift. We denote the input $\boldsymbol{x} = x_1x_2 \cdots x_n$, where $\boldsymbol{x} \in \{0, 1\}^n$. Furthermore, there are m output qubits denoted by $|0\rangle_1, |0\rangle_2, \ldots, |0\rangle_m$. The value of m is determined by the choice of α.

Theorem 4.2 *There is a quantum circuit C_n capable of implementing the n-bit ReLU function with a constant T-depth of 4 without ancillary qubits. Furthermore, the circuit depth is $O(\log n)$, and the circuit size is $O(n)$.*

The quantum circuit is constructed in two parts. The first part resembles the circuit in Theorem 4.2: We start by applying an X gate to $|x_1\rangle$, followed by $n-1$ Toffoli gates. The i-th Toffoli gate has $|x_1\rangle$ as the first control qubit, $|x_{i+1}\rangle$ as the

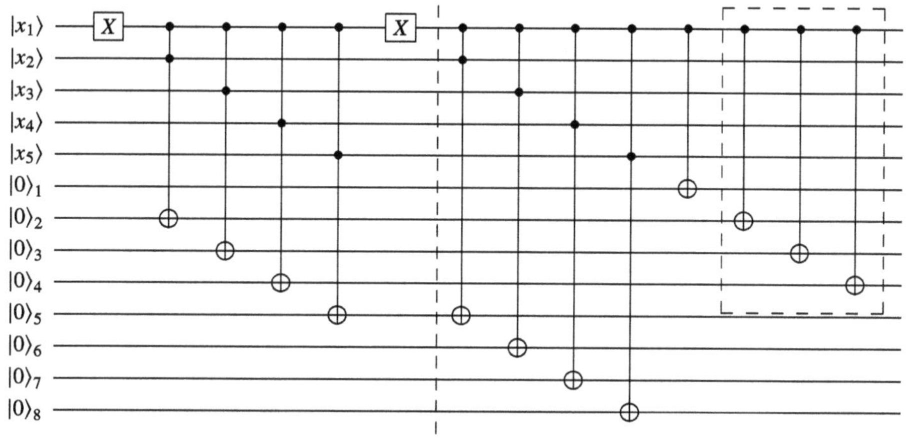

Fig. 4.2 The quantum circuit for the 5-bit Leaky ReLU function. Here, $|x_i\rangle$ represents input qubits, and $|0\rangle_j$ represents output qubits

second control qubit, and $|0\rangle_{i+1}$ as the target qubit. Finally, we apply another X gate to $|x_1\rangle$. This first part of the circuit copies the state of $|x_i\rangle$ to $|0\rangle_i$ for $i = 2, 3, \ldots, n$ when $x_1 = 0$, corresponding to the case where $x \geq 0$, and we should output x itself.

The second part of the circuit is constructed as follows: We apply n-1 Toffoli gates, where the i-th Toffoli gate has $|x_1\rangle$ as the first control qubit, $|x_{i+1}\rangle$ as the second control qubit, and $|0\rangle_{i+k}$ as the target qubit, with k=1 + $\log \frac{1}{\alpha}$. Then, we apply a CNOT gate with $|x_1\rangle$ as the control qubit and $|0\rangle_1$ as the target qubit. If we use the True form, the circuit is complete. However, if we use the 2's complement, we need to apply $k - 1$ additional CNOT gates, where the control qubit is $|x_1\rangle$ and the target qubits are $|0\rangle_2, |0\rangle_3, \ldots, |0\rangle_k$, respectively. An example of a quantum circuit implementing the 5-bit Leaky ReLU function with $\alpha = 0.125$ is depicted in Fig. 4.2. In this figure, the circuit constructed in the first part is shown on the left side of the barrier. When using the true form, the gates inside the dashed box need to be omitted.

4.2.4 Results and Discussions

To verify the correctness of the circuit, let us examine the value of x_1.

When x_1=0, all quantum gates in the second part can be omitted. The circuit's correctness follows from the same proof as in Theorem 4.2.

When x_1=1, all quantum gates in the first part can be omitted. The circuit will then copy $|x_i\rangle$ to $|0\rangle_{i+k}$ for i=2, 3, ..., n, where k=1 + $\log \frac{1}{\alpha}$. Subsequently, we copy $|x_1\rangle$ to $|0\rangle_1$. If we use the true form, we already obtain αx. However, if we use 2's complement, we need to flip $x_2, \ldots, x_k$. Thus, we successfully store αx on the output qubits.

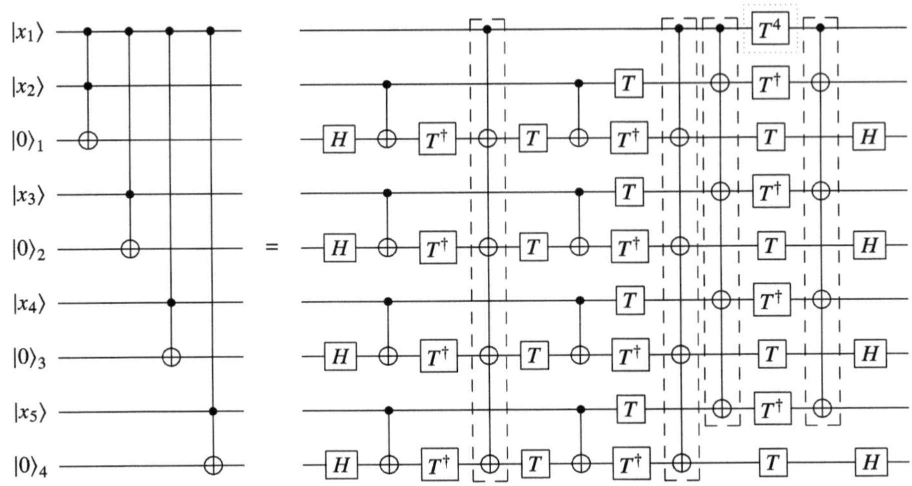

Fig. 4.3 The quantum circuit for the four shared-control Toffoli gates

To compute the T-depth of our quantum circuit, we observe that only the Toffoli gates contribute to it. By utilizing the method illustrated in Fig. 4.3 [68], we find that the T-depth of our circuit is 8. Similarly, we can determine that the circuit depth and size are both $O(\log n)$ and $O(n)$, respectively.

4.2.5 Section Summary

This work focuses on implementing activation functions with lower T-depth quantum circuits, yielding significant breakthroughs. Specifically, we achieve constant T-depth circuits for Leaky ReLU function, regardless of the number of input qubits, n. We have balanced the number of input qubits against implementation precision and the trade-off between circuit T-depth and ancilla count to enhance the applicability of our results. Our method demonstrates advantages through a specific analysis of T-depth comparisons.

Looking ahead, we anticipate leveraging our activation function implementations in the implementation of quantum neural networks, thereby paving the way for future research directions in this domain.

4.3 Applications in Quantum Cryptanalysis

In our exploration of quantum cryptanalysis, we conducted a work titled "New Results in Quantum Analysis of LED: Featuring One and Two Oracle Attacks."

This analysis incorporates several efficient implementations of quantum arithmetic operations, such as MixColumn.

4.3.1 Introduction

The Internet of Things (IoT) is revolutionizing the way devices communicate and operate through interconnected devices, resulting in a large number of applications across a variety of domains [31, 123]. Nonetheless, IoT security remains a significant concern due to the large number of devices and their limited resources. Lightweight cryptography aims to address these limitations, ensuring secure communication while maintaining efficiency in terms of power, speed, and hardware resources. However, with the development of quantum computing, quantum attacks such as Grover's algorithm pose substantial risks to encryption systems, thereby significantly impacting the security of lightweight cryptography in the quantum computing era.

Therefore, it is crucial to explore and analyze lightweight ciphers in the context of quantum attacks. Several relevant studies have already been conducted [104, 106], focusing on lightweight block ciphers such as PRESENT, GIFT, and SPECK. Moreover, we have seen recent improvements on the attack on AES, namely [129, 192].

Among various lightweight block ciphers, the LED cipher stands out due to its compact hardware design and minimal silicon footprint. Given these properties, this work focuses on the quantum implementation and quantum attack of the LED cipher. Specifically, we propose an efficient quantum circuit for the LED cipher, minimizing the required qubits, quantum gates, and circuit depth. Furthermore, we perform Grover's attack and Search with Two Oracles (STO) attack on the proposed quantum LED cipher and estimate the quantum resources required for these attack oracles. This work also addresses several issues found in the quantum LED cipher proposed by Song et al. [197]. By comparing it with the corrected version of their implementation, it is evident that our design provides a more comprehensive and efficient solution. Additionally, this research addresses modes of operation in the context of quantum attack. In classical cryptography, modes of operation such as ECB, CBC, OFB, and CFB have been widely discussed and standardized since their introduction in FIPS 81 back in 1981. However, in quantum cryptanalysis, while quantum attacks on block ciphers have been extensively explored, modes of operation have not been considered thus far. To fill this critical gap, this work proposes frameworks for ECB, CBC, and CFB modes based on the quantum LED cipher. Furthermore, we provide a thorough analysis of the quantum attack costs associated with different modes of operation.

4.3.1.1 Contributions and Novelty

- **Efficient Quantum Circuits for LED Block Cipher.** We present a novel approach to implementing the LED block cipher as a quantum circuit. The proposed approach optimizes the quantum circuit of the LED cipher by exploring efficient subcircuit structures, thereby enhancing the overall efficiency.
- **Quantum Resource Analysis and Comparison.** We evaluate the performance of our design using the ProjectQ framework [201]. Our analysis includes a detailed comparison of the required quantum resources, such as qubits and circuit depth, with respect to other block cipher implementations in the literature. Compared to previous works, our implementation achieves a more efficient quantum circuit for the LED cipher.
- **Quantum Cost Estimation of Grover's Attack and STO Attack.** In this chapter, Grover's attack is performed on the proposed quantum LED cipher. Moreover, as referenced in [43, 116], we also conduct the STO attack, which is an improvement over Grover's attack when the state size is less than the key size for a block cipher. For both types of attacks, we estimate the quantum resource requirements and provide a comprehensive evaluation of their efficiency.
- **Modes of Operation for Quantum Block Ciphers.** This work explores the modes of operation in quantum attacks, proposing frameworks for ECB, CBC, and CFB modes based on the proposed quantum LED cipher. Besides, it also provides a thorough analysis of Grover's attack's effectiveness across these modes.

4.3.2 Background

4.3.2.1 LED Block Cipher

In 2011, Guo et al. [75] introduced the LED block cipher, a compact symmetric key block cipher specifically designed for efficient encryption in resource-constrained environments. LED operates on 64-bit blocks and supports key sizes of 64 and 128 bits. The cipher employs 32 rounds for the 64-bit key version and 48 rounds for the 128-bit key version. Specifically, before the first round, the LED cipher performs an initial AddRoundKey operation. Subsequently, the AddRoundKey operation is executed every four rounds. Each round consists of four steps performed sequentially: AddConstants, SubCells, ShiftRows, and MixColumns. Besides, it is necessary to highlight that performing one MixColumnsSerial requires executing four MixColumn operations. An overview of the LED encryption process is shown in Fig. 4.4, with a detailed discussion of each component as follows:

- **KeySchedule.** LED employs a straightforward key schedule. For LED-64, the same 64-bit user key K is used directly in each round. While for LED-128, the

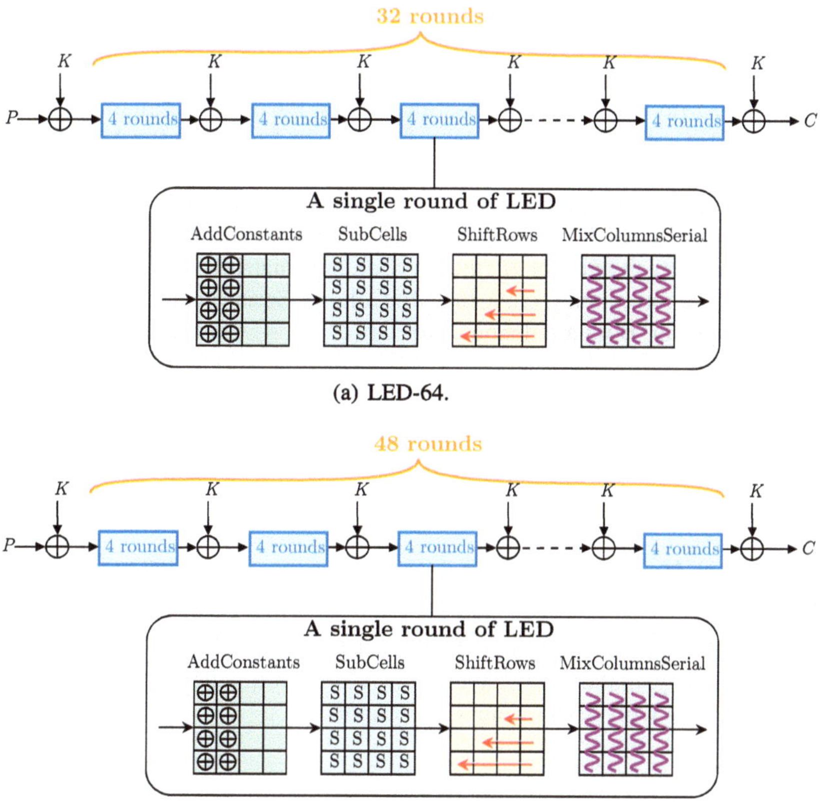

(a) LED-64.

(b) LED-128.

Fig. 4.4 Schematics for LED block cipher

Table 4.2 Round constants for the LED block cipher

Rounds	Constants
1–24	01, 03, 07, 0F, 1F, 3E, 3D, 3B, 37, 2F, 1E, 3C, 39, 33, 27, 0E, 1D, 3A, 35, 2B, 16, 2C, 18, 30
25–48	21, 02, 05, 0B, 17, 2E, 1C, 38, 31, 23, 06, 0D, 1B, 36, 2D, 1A, 34, 29, 12, 24, 08, 11, 22, 04

128-bit user key is divided into two subkeys ($K = K0||K1$), where the key in each round is alternately set to equal the left part $K0$ and the right part $K1$ of K.

- **AddRoundKey.** As defined by the previously discussed **KeySchedule**, the appropriate round keys are generated based on the key size to ensure proper encryption. In this step, the 64-bit block is exclusive-ored with the corresponding 64-bit round key.
- **AddConstants.** The round constants are detailed in Table 4.2, which presents the constants $(rc_5, rc_4, rc_3, rc_2, rc_1, rc_0)$ encoded as byte values for each round. Particularly, rc_0 represents the least significant bit.

- **SubCells.** The LED cipher reuses the S-box from the PRESENT block cipher [21] $(C, 5, 6, B, 9, 0, A, D, 3, E, F, 8, 4, 7, 1, 2)$.
- **ShiftRows.** This operation involves cyclically left shifting the bytes in $i-th$ row of the state array by i cell positions. Specifically, the $0th$ row remains unchanged, while the $1st$ row is shifted left by 4 bits, the $2nd$ row is shifted left by 8 bits, and the $3rd$ row is shifted left by 12 bits.
- **MixColumns.** This operation processes each column of the state array as a column vector, which is then replaced by a new vector obtained by post-multiplying it by the matrix M (the MixColumns matrix). This matrix is actually obtained by another matrix, A (the MixColumnsSerial matrix), by raising it to the fourth power. Both are defined over $\mathrm{GF}(2^4)/x^4 + x + 1$ and are given by

$$M = \begin{bmatrix} 4 & 1 & 2 & 2 \\ 8 & 6 & 5 & 6 \\ B & E & A & 9 \\ 2 & 2 & F & B \end{bmatrix} \qquad A = \begin{bmatrix} 0 & 1 & 0 & 0 \\ 0 & 0 & 1 & 0 \\ 0 & 0 & 0 & 1 \\ 4 & 1 & 2 & 2 \end{bmatrix}.$$

  Note that $M = A^4$ and is MDS.[4] The authors recommended to implement M by successively applying the circuit of A four times for low resource utilization.

In brief, the prudent design of LED strikes a balance between security and resource efficiency, making this cipher particularly well-suited for IoT devices and other applications with strict resource constraints.

4.3.2.2 Quantum Gates for Cipher Implementations

As shown in Fig. 4.5, various quantum gates play crucial roles in the implementation of quantum block ciphers.

- **X Gate.** This quantum gate performs a bit-flip operation, inverting the state of a qubit from $|x\rangle$ to $|\neg x\rangle$ and vice versa.
- **CNOT Gate.** This gate is a two-qubit gate that flips the state of the target qubit if the control qubit is in the state $|1\rangle$.
- **Toffoli Gate.** This operation, also known as the controlled-controlled-NOT (CCNOT) gate, flips the state of the target qubit only when both control qubits are in the state $|1\rangle$. It is noteworthy that this quantum gate can be decomposed into basic quantum gates such as X, CNOT, H, and T gates [5] in various ways. Consequently, the implementation cost depends on the specific decomposition method.
- **SWAP Gate.** This is a two-qubit gate that exchanges the states of two qubits. As illustrated in Fig. 4.5d, the operation SWAP(x, y) results in (y, x).

[4] Also, one may note that $A = M^{64}$.

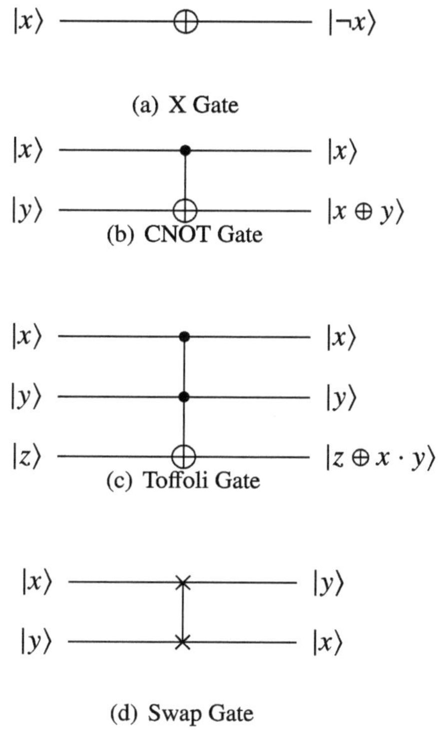

Fig. 4.5 Common quantum gates required for cipher implementations

In brief, these four quantum gates form the essential building blocks required for the implementation of quantum block ciphers.

4.3.2.3 Grover's Attack

In quantum cryptanalysis, Grover's attack can be utilized to perform a quantum key search, significantly reducing the time complexity of finding the correct encryption key.

Before delving into Grover's attack, it is essential to first understand Grover's algorithm. This famous quantum algorithm provides a quadratic speedup over classical algorithms for unstructured search problem. Specifically, Grover's algorithm can locate the solution to a problem within an unsorted database of N items using approximately $O(\sqrt{N})$ queries, compared to the $O(N)$ queries required by classical brute-force method.

Grover's attack is based on the principles of Grover's algorithm. Figure 4.6 illustrates the quantum oracles used in this attack, clearly demonstrating how it directly borrows and adapts the structure of Grover's algorithm to achieve its key search objective. Specifically, this attack begins by applying Hadamard gates to the key qubits, which creates superposition states. Next, the specific encryption

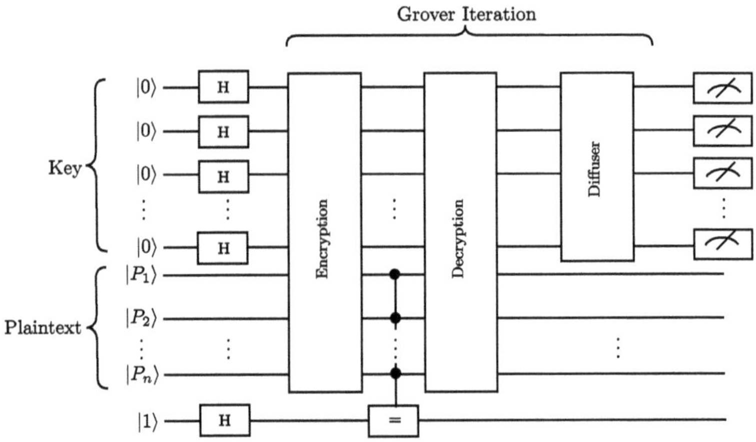

Fig. 4.6 Quantum oracles for Grover's algorithm and attack

quantum circuit uses these states to encrypt the plaintext. If the generated ciphertext matches the known ciphertext, this oracle then inverts the sign of the corresponding key state. Following this, the diffusion operator is applied to amplify the amplitude of the potential solution key. The combination of the oracle and diffusion operator is repeated multiple times to further enhance the amplitude of the correct key. Finally, measuring the key qubits reveals the most probable solution.

In this chapter, we perform a Grover's attack on the proposed LED cipher, along with an estimation of the quantum resources required for the corresponding attack oracle.

4.3.2.4 Improvement over Grover: Search with Two Oracles Attack

In this subsection, we introduce the search with two Oracles (STO) attack as an enhancement of Grover's algorithm.

Initially proposed by Kimmel et al. [116], the STO attack is able to reduce the overall cost of quantum search procedures compared to directly applying Grover's algorithm. This attack utilizes two quantum oracles: The first oracle, O_γ, is relatively inexpensive and marks both the M target items and a number of false positives; the second, O_χ, is more expensive but accurately identifies the M target items. Particularly, O_χ is identical to the one used in Grover's algorithm. For STO attack, the number of queries required remains $O(\sqrt{2^k/M})$ as that for Grover's attack, where k denotes the key size. In brief, this attack enhances the overall efficiency by balancing the costs of O_γ and O_χ with the number of false positives. By carefully designing these oracles and managing their associated costs, the search process can be significantly optimized.

The STO attack we implement in this work follows a similar methodology to that described in [43]. In particular, our construction of the oracle O_{χ} employs a serial-oracle design pattern, as suggested in the same reference. However, no further optimizations were made to O_{γ} in this work, as our proposed quantum LED cipher is already highly efficient and lightweight.

In subsequent content, we provide a thorough analysis of the quantum resources required for the proposed STO attack on LED cipher.

4.3.3 Methodology

4.3.3.1 Method 1: Efficient Quantum Implementation of LED Cipher

In this section, we present an efficient quantum implementation of the LED cipher, focusing on minimizing the overall cost. This cipher consists of several submodules, which are arranged into multiple rounds of encryption. Accordingly, we will detail the efficient implementations for the five key submodules of quantum LED cipher in the subsequent content.

Submodule 1: AddRoundKey
The quantum AddRoundKey operation is implemented using only CNOT gates, resulting in a circuit with a depth of one. This straightforward operation is simpler than other components like the S-box and MixColumn, as it follows a generic implementation method. Specifically, in the LED cipher, a 64-qubit round key is XORed with the intermediate state using 64 CNOT gates, as illustrated in ProjectQ Code 4.3.1.

Code 4.3.1 AddRoundKey of quantum LED implementation

```
def KeyAddition(eng, state, key):
    for i in range(64):
        CNOT | (key[i], state[i])
```

Submodule 2: AddConstants
Similar to the AddRoundKey operation, the quantum AddConstants operation is also based on a generic method. Since the round constant in the LED cipher is a fixed value, as detailed in Table 4.2, this quantum implementation only involves applying X gates to the state qubits where the corresponding bits in the round constant are set to 1. The specific ProjectQ implementation details are provided in Code 4.3.2.

Code 4.3.2 AddConstants of quantum LED implementation

```
def Round_constant_XOR(eng, state, round_constant):
    for i in range(64):
        if (round_constant >> i & 1):
            X | state[i]
```

Submodule 3: SubCells

The SubCells operation applies a nonlinear substitution using S-box. As mentioned in Sect. 4.3.2.1, the LED cipher reuses the same S-box as PRESENT [21]. The specific coordinate functions of this S-box are as described as follows:

$$y_0 = x_0 \oplus x_1x_2 \oplus x_2 \oplus x_3$$
$$y_1 = x_0x_1x_2 \oplus x_0x_1x_3 \oplus x_0x_2x_3 \oplus x_1x_3 \oplus x_1 \oplus x_2x_3 \oplus x_3$$
$$y_2 = x_0x_1x_3 \oplus x_0x_1 \oplus x_0x_2x_3 \oplus x_0x_3 \oplus x_1x_3 \oplus x_2 \oplus x_3 \oplus 1$$
$$y_3 = x_0x_1x_2 \oplus x_0x_1x_3 \oplus x_0x_2x_3 \oplus x_0 \oplus x_1x_2 \oplus x_1 \oplus x_3 \oplus 1.$$

For the quantum implementation, the design of an efficient quantum S-box is crucial, as it directly impacts the performance of this operation. Given the inherent complexity and the numerous possible approaches for implementing a quantum LED S-box, we have explored a range of strategies to identify the most efficient design. These strategies are outlined in detail below:

- **Naïve.** This approach directly utilizes the S-box coordinate functions to design the quantum circuit. While straightforward, this method does not optimize for quantum resources, resulting in higher quantum costs.
- **LIGHTER-R.** LIGHTER-R [42] is a tool designed for implementing S-boxes using reversible logic libraries, specifically targeted for 4-bit S-boxes. Using this tool, we generated a quantum implementation of the LED S-box, which is illustrated in Fig. 4.7b.
- **DORCIS.** DORCIS [33] is a tool that optimizes quantum implementations for arbitrary 3- and 4-bit S-boxes, extending the capabilities of the LIGHTER-R. Unlike LIGHTER-R, which only handles 4-bit S-boxes and operates at a top level using Toffoli gates, DORCIS incorporates quantum decomposition with Clifford and T gates, optimizing both quantum depth and T-depth. In this chapter, we employed DORCIS to generate a quantum implementation of the LED S-box, as shown in Fig. 4.7c.
- **S-box α.** In the paper [38], Courtois et al. optimized various small digital circuits, including LED S-box. It is shown in Fig. 4.7d.
- **S-box β.** In 2024, Feng et al. [52] proposed optimized implementations of lightweight cryptographic S-boxes using SAT solvers, which significantly enhance the overall cryptographic performance. Here, we implement their optimized LED S-box into quantum circuit, as shown in Fig. 4.7e.

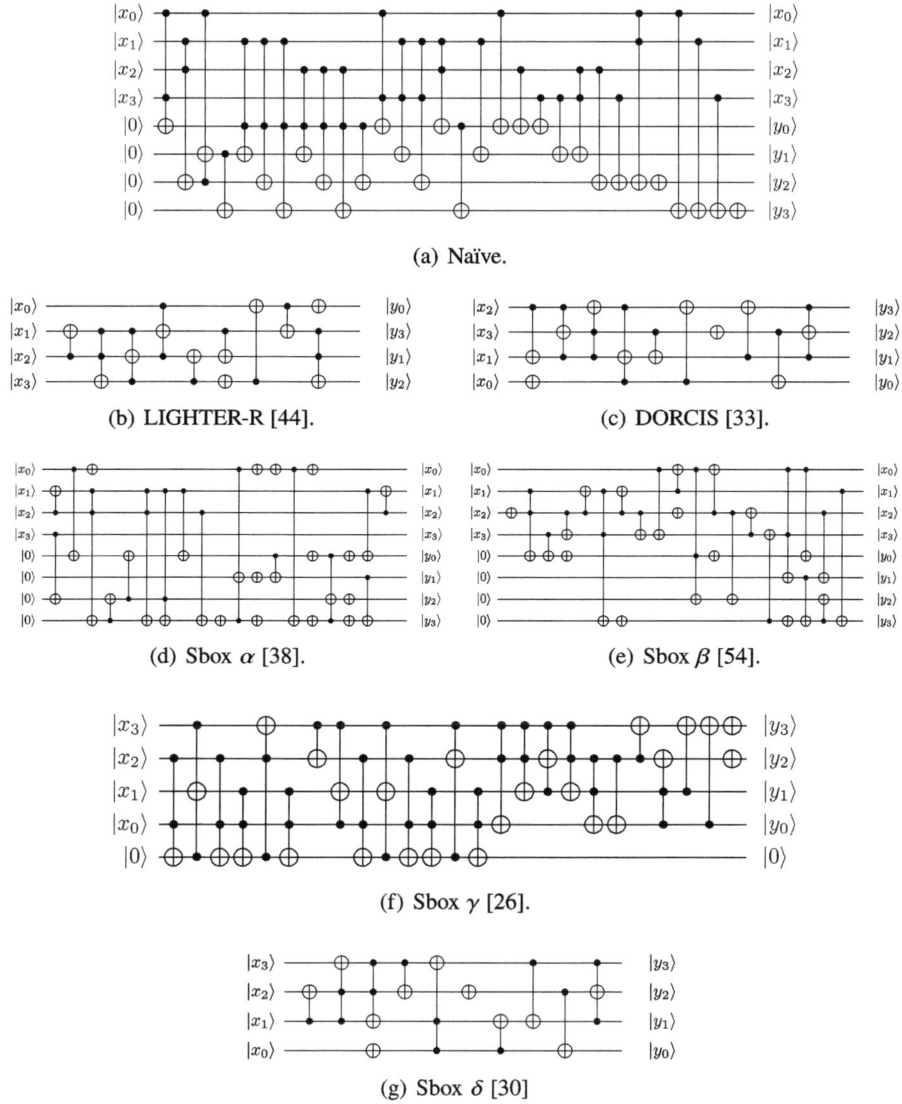

(a) Naïve.

(b) LIGHTER-R [44].

(c) DORCIS [33].

(d) Sbox α [38].

(e) Sbox β [54].

(f) Sbox γ [26].

(g) Sbox δ [30]

Fig. 4.7 Quantum implementations of LED S-box

- **S-box** γ. In the paper by Cai et al. [26], a quantum S-box for the LED cipher is proposed that utilizes five qubits (i.e., 1 ancilla/garbage qubit). This design is illustrated in Fig. 4.7f.
- **S-box** δ. In 2024, Chen et al. [30] proposed a SAT-based model that optimizes quantum circuits incorporating three metrics. Based on this model, we designed a more compact quantum circuit for LED S-box, as illustrated in Fig. 4.7g.

In Table 4.3, the quantum costs associated with each approach are presented. It is obvious that both the S-box γ by Chen et al. and the DORCIS-generated S-box

Table 4.3 Quantum resource requirement for LED S-Box. Here, "1qClifford" refers to the Clifford gate operating on a single qubit, such as X gate

Method	#1qCliff	#CNOT	#Toffoli	#T	#Qubits	Circuit depth	#Toffoli depth
Naïve (Fig. 4.7a)	2	13	15	105	8	27	14
LIGHTER-R [42] (Fig. 4.7b)	2	5	4	28	4	9	4
DORCIS [33] (Fig. 4.7c)[a]	2	5	4	28	4	8	4
S-box α [38] (Fig. 4.7d)	12	12	5	35	8	16	5
S-box β [52] (Fig. 4.7e)	6	16	4	28	8	19	4
S-box γ [26] (Fig. 4.7f)	7	5	19	133	5	24	19
S-box δ [30] (Fig. 4.7g)	2	5	4	28	4	8	4

a: Used in this work

exhibit the highest efficiency, with a circuit depth of 8 and a Toffoli depth of 4. Since the two designs are very similar, either one could be chosen. In this chapter, we decided to use the DORCIS-generated implementation for our quantum S-box.

Submodule 4: ShiftRows

The ShiftRows operation, which performs circular shifts, can be implemented using quantum swap gates, also known as "physical swap" approach. Alternatively, this operation can be achieved using the "logical swap" approach, which involves rearranging the indices of the qubits logically without employing additional quantum swap gates. As one can see, the second approach is more resource-efficient, as it avoids the requirement for extra quantum resources. Therefore, in this chapter, we adopt the "logical swap" approach. Details of the implementation are provided in Code 4.3.3.

Code 4.3.3 ShiftRows of quantum LED implementation

```
def ShiftRows(eng, state):
    newstate=[]
    #1st row-left rotate 4
    newstate[0:4]=state[4:8]
    newstate[4:8]=state[8:12]
    newstate[8:12]=state[12:16]
    newstate[12:16]=state[0:4]
    #2nd row-left rotate 8
    newstate[16:20]=state[24:28]
    newstate[20:24]=state[28:32]
    newstate[24:28]=state[16:20]
    newstate[28:32]=state[20:24]
    #3rd row-left rotate 12
    newstate[32:36]=state[44:48]
    newstate[36:40]=state[32:36]
    newstate[40:44]=state[36:40]
    newstate[44:48]=state[40:44]
    #0th row-left rotate 0
    newstate[48:64]=state[48:64]
    return newstate
```

Submodule 5: MixColumn

Based on the specification of LED (see Sect. 4.3.2.1), the MixColumnsSerial matrix (A) and the MixColumns matrix can be given in binary as follows:

$$
M = \begin{bmatrix}
0&0&1&0&1&0&0&0&0&1&0&0&0&1&0&0 \\
1&0&0&1&0&1&0&0&0&0&1&0&0&0&1&0 \\
1&1&0&0&0&0&1&0&1&0&0&1&1&0&0&1 \\
0&1&0&0&0&0&0&1&1&0&0&0&1&0&0&0 \\
1&0&0&1&0&1&1&0&1&0&1&0&0&1&1&0 \\
1&1&0&0&1&0&1&1&1&1&0&1&1&0&1&1 \\
0&1&1&0&0&1&0&1&1&1&1&0&0&1&0&1 \\
0&0&1&0&1&1&0&0&0&1&0&1&1&1&0&0 \\
0&1&0&1&1&1&1&1&1&1&0&1&0&0&0&1 \\
1&0&1&0&0&1&1&1&1&1&1&0&1&0&0&0 \\
1&1&0&1&0&0&1&1&1&1&1&1&0&1&0&0 \\
1&0&1&1&1&1&1&0&1&0&1&0&0&0&1&1 \\
0&1&0&0&0&1&0&0&0&1&1&1&0&1&0&1 \\
0&0&1&0&0&0&1&0&0&0&1&1&1&0&1&0 \\
1&0&0&1&1&0&0&1&0&0&0&1&1&1&0&1 \\
1&0&0&0&1&0&0&0&1&1&1&1&1&0&1&1
\end{bmatrix}
A = \begin{bmatrix}
0&0&0&0&1&0&0&0&0&0&0&0&0&0&0&0 \\
0&0&0&0&0&1&0&0&0&0&0&0&0&0&0&0 \\
0&0&0&0&0&0&1&0&0&0&0&0&0&0&0&0 \\
0&0&0&0&0&0&0&1&0&0&0&0&0&0&0&0 \\
0&0&0&0&0&0&0&0&1&0&0&0&0&0&0&0 \\
0&0&0&0&0&0&0&0&0&1&0&0&0&0&0&0 \\
0&0&0&0&0&0&0&0&0&0&1&0&0&0&0&0 \\
0&0&0&0&0&0&0&0&0&0&0&1&0&0&0&0 \\
0&0&0&0&0&0&0&0&0&0&0&0&1&0&0&0 \\
0&0&0&0&0&0&0&0&0&0&0&0&0&1&0&0 \\
0&0&0&0&0&0&0&0&0&0&0&0&0&0&1&0 \\
0&0&0&0&0&0&0&0&0&0&0&0&0&0&0&1 \\
0&0&1&0&1&0&0&0&0&1&0&0&0&1&0&0 \\
1&0&0&1&0&1&0&0&0&0&1&0&0&0&1&0 \\
1&1&0&0&0&0&1&0&1&0&0&1&1&0&0&1 \\
0&1&0&0&0&0&0&1&1&0&0&0&1&0&0&0
\end{bmatrix}
$$

Here, we explored several methods to determine the most efficient design, which are detailed as follows:

- **Naïve (Out-of-place).** The naïve approach involves using ancilla qubits to compute the new state of a qubit while preserving its previous state in a backup qubit, which is essential for the calculation of new state of two other qubits. Following this approach, Song et al. [197] proposed the naïve implementation of the MixColumn operation in the quantum LED cipher. As shown in Table 4.4, this implementation requires 32 qubits and 108 CNOT gates, with a quantum depth of 10.
- **PLU Factorization.** PLU factorization decomposes a given binary matrix into a permutation matrix, a lower triangular matrix, and an upper triangular matrix. This factorization can also be applied to achieve an efficient in-place quantum implementation for MixColumn operation. By following the method detailed in [216], we use Sage[5] to obtain an in-place quantum implementation through PLU factorization.
- **Gauss-Jordan Elimination.** Gauss-Jordan elimination is used to factorize any binary matrix through elementary operations, which correspond to CNOT and SWAP gates in quantum circuits. By applying this method, we achieve an in-

[5] https://doc.sagemath.org/html/en/reference/matrices/sage/matrix/matrix2.html

Table 4.4 Quantum resource requirement for LED MixColumn

Method	#CNOT	#SWAP	Circuit depth
Out-of-place (32 qubits)			
Naïve (used in [197])	108	0	10
In-place (16 qubits)			
PLU	103	8	62
Gauss-Jordan	103	8	52
XZLBZ [236][a]	44	16	16
YWSZZ [244]	47	16	10

a: Used in this work

place quantum implementation that utilizes 16 qubits with 103 CNOT gates and 8 SWAP gates, with a quantum depth of 52.

- **XZLBZ.** The XZLBZ algorithm [236] offered an innovative approach for the in-place quantum implementation of binary matrices. At that time, it was the first tool to efficiently implement a given linear layer. The most notable result from this work was to find an in-place implementation of the AES MixColumn with 92 CNOT gates and 30 quantum depth (although they did not optimize for quantum depth).
- **YWSZZ.** This is a very recent paper [244] where the authors managed to find an in-place implementation of the AES MixColumn matrix with 91 CNOT gates and 13 quantum depth. This is where the record stands till date, to the best of our knowledge. In our case though, it seems that XZLBZ (44) outperformed YWSZZ (47) in terms of CNOT gates.

As shown in Table 4.4, we summarize the quantum costs for the various approaches discussed before. Obviously, the XZLBZ implementation demonstrates the highest efficiency, achieving 44 CNOT gates with 16 qubits and a circuit depth of 16. Hence, we chose this implementation as our quantum MixColumn.

The matrix A requires 14 CNOT gates in the optimal in-place implementation. This was found using the MILP tool of [11]. Thus, it would require $4 \times 14 = 52$ CNOT gates to implement M by using this matrix, and hence were not considered here. Using the same tool, we also found that the minimum number of CNOT gates that are required to implement M in-place is at least 15.[6]

Overall Architecture

Using the presented modules AddConstants, SubCells, ShiftRows, and MixColumns, we construct the architecture for the LED quantum circuit. Even at the architectural level, the in-place design is implemented, where the ciphertext is computed directly on the input qubits (i.e., plaintext) without allocating additional qubits.

[6] The program took a long time since it searched for an implementation with 15 CNOT gates, yet there was no outcome.

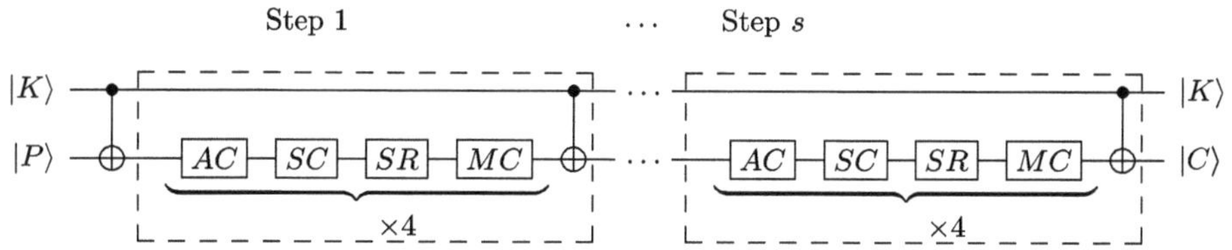

Fig. 4.8 The in-place architecture of the LED quantum circuit. AC, SC, SR, and MC represent AddConstants, SubCells, ShiftRows, and MixColumns, respectively

Figure 4.8 shows the in-place architecture of the LED quantum circuit. The only differences between LED-64 and LED-128 are the number of steps (s) and the AddRoundKey operation, neither of which affects the overall architecture. Thus, the in-place architecture shown in Fig. 4.8 applies equally to both the LED-64 and LED-128 quantum circuits ($s = 8$ and $s = 12$, respectively).

4.3.3.2 Method 2: Consideration for Modes of Operation

In classical cryptography, modes of operation were introduced to enhance security when using block ciphers to encrypt long messages. However, in the field of quantum cryptanalysis, this topic has not received much attention. To address this gap, this work proposes several frameworks for modes of operation based on the proposed quantum LED circuit. In this chapter, we focus on three widely utilized modes of operation: Electronic Codebook (ECB), Cipher Block Chaining (CBC), and Cipher Feedback (CFB). The detailed construction methods for these modes are illustrated in Fig. 4.9.

The following subsections will provide detailed implementations of the quantum frameworks for ECB, CBC, and CFB modes. For clarity and conciseness, the simplified representation of the proposed quantum LED implementation, as shown in Fig. 4.10, will be utilized throughout the remaining part of this section.

Mode 1: Electronic Codebook Mode (ECB)
As shown in Fig. 4.9a, in ECB mode, each plaintext block is independently encrypted using the same block cipher, resulting in a corresponding ciphertext block. This mode is straightforward, as it directly applies the quantum cipher to each block without any interdependencies between them. However, the absence of inter-block dependencies in ECB can make it very vulnerable to various attacks.

The quantum framework for ECB mode is depicted in Fig. 4.11, where each plaintext block is processed independently using the quantum LED circuit.

Mode 2: Cipher Block Chaining Mode (CBC)
In CBC mode, interdependencies between plaintext blocks are established by XORing each plaintext block with the ciphertext of the previous block before

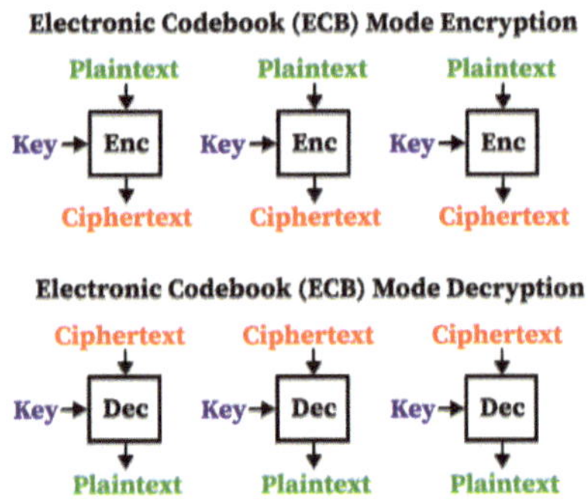

(a) ECB

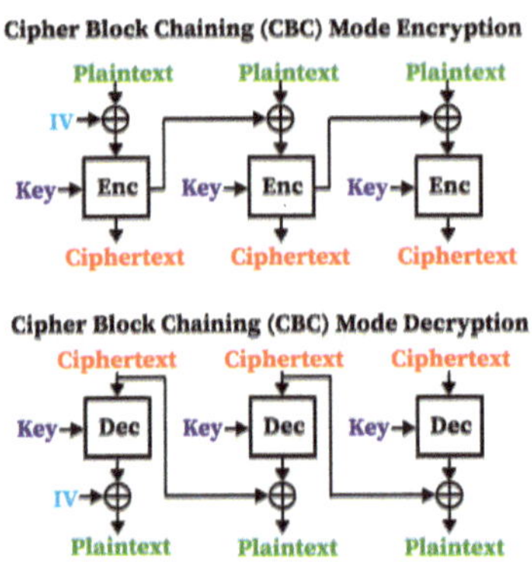

(b) CBC

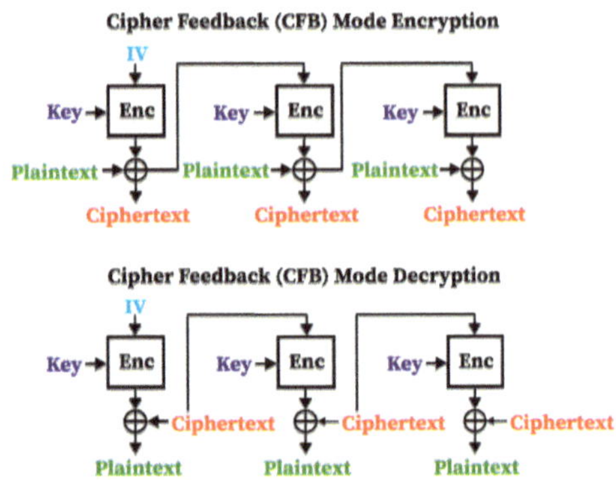

(c) CFB

Fig. 4.9 Typical modes of operations considered in this work

Fig. 4.10 Simplified representation of LED in quantum

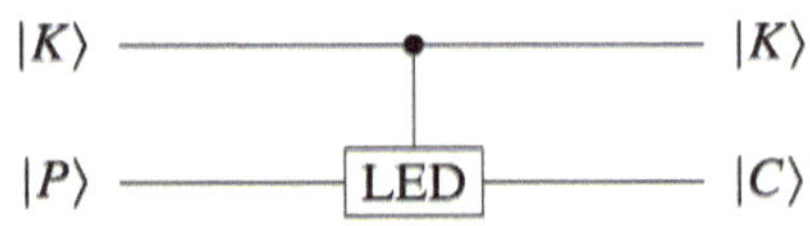

encryption. For the first plaintext block, the XOR operation is performed with an initialization vector (IV), which is typically a random or pseudorandom value. This chaining mechanism ensures that identical plaintext blocks produce different ciphertexts, thereby enhancing security compared to ECB mode.

The quantum framework for CBC mode is shown in Fig. 4.12, illustrating the process of XORing each plaintext block with the previous ciphertext block, followed by encryption using the quantum LED cipher.

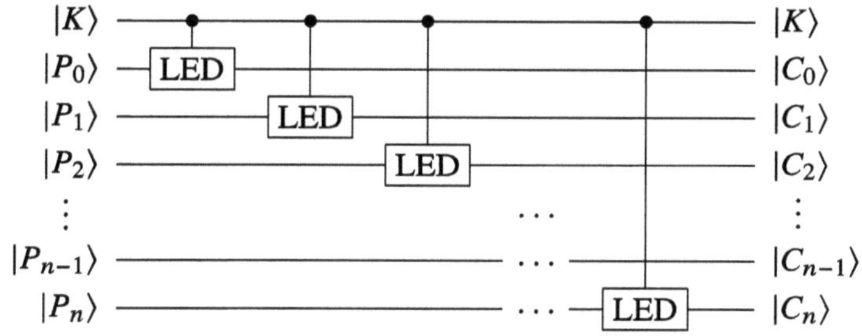

Fig. 4.11 Framework for ECB mode based on the proposed quantum LED. In this diagram, K denotes the key, P denotes plaintext, and C denotes ciphertext, with all ancilla qubits in the quantum LED cipher omitted for conciseness

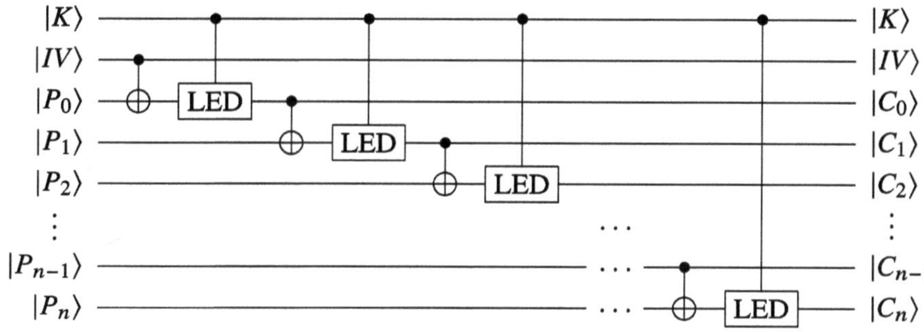

Fig. 4.12 Framework for CBC mode based on the proposed quantum LED. Here, P denotes plaintext, C denotes ciphertext, IV denotes initialization vector, with all ancilla qubits in the quantum LED cipher omitted for conciseness

Mode 3: Cipher Feedback Mode (CFB)
The CFB mode operates similarly to a stream cipher, converting a block cipher into a stream cipher. Particularly, this mode is suitable for real-time data encryption by allowing for the partial encryption of data streams. In CFB, the previous ciphertext block is encrypted using the block cipher, and the resulting output is then XORed with the current plaintext block to generate the new ciphertext block.

The quantum framework for implementing CFB mode is provided in Fig. 4.13, where the feedback mechanism and sequential encryption process are clearly illustrated. Although the proposed framework generates garbage qubits, all of them will be cleaned to $|0\rangle$ through the subsequent uncomputation step in Grover's attack.

In this part, we have proposed frameworks for ECB, CBC, and CFB modes based on our quantum LED implementations. Subsequently, Grover's attack estimation will be performed on the LED cipher under these proposed modes of operation frameworks to evaluate their effectiveness.

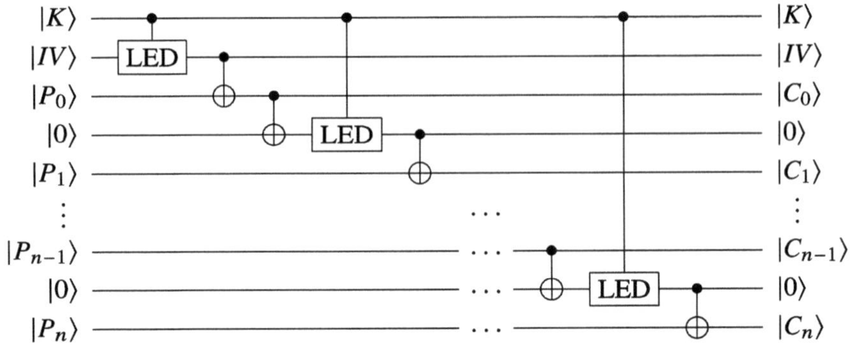

Fig. 4.13 Framework for CFB mode based on the proposed quantum LED. Here, P denotes plaintext, C denotes ciphertext, IV denotes initialization vector, with all ancilla qubits in the quantum LED cipher omitted for conciseness

4.3.4 Results and Discussions

Here, we conduct a thorough analysis of all the experimental results. To begin with, the issues in the previous quantum LED implementation proposed by Song et al. [197] are revised, thereby establishing a more accurate benchmark for subsequent comparisons. Following this, we compare the quantum resources required to implement the LED cipher with those needed for other block ciphers. Besides, this work provides the quantum attack resources based on the proposed quantum LED implementations, including Grover's attack oracles and STO attack oracles. At the end of this part, the effectiveness of Grover's attack oracles on various quantum modes of operation is carefully estimated, with the proposed quantum LED cipher serving as the underlying block cipher.

4.3.4.1 Revision of the Previous Quantum LED Implementations

In the previous work [197], Song et al. also proposed a quantum implementation of the LED cipher. However, this work contains an issue in estimating the required number of qubits for the quantum implementation of LED cipher.[7] Here, we analyze the issue and provide a correction to ensure a more accurate comparison in subsequent content.

As shown in Table 4.4, the authors proposed an out-of-place naïve MixColumn. To perform one MixColumnsSerial, four MixColumn operations are required, with each MixColumn requiring 16 newly allocated output qubits, totaling 64 qubits for a single MixColumnsSerial. Since LED-64 consists of 32 rounds, each applying

[7] We had personal communication with the authors of this paper in September 2024. They acknowledged this issue and granted us permission to revise it in our work.

Table 4.5 Revised benchmark for the quantum implementation of LED-64

Quantum circuit	#Qubits	#Toffoli	#CNOT	#X	Circuit depth
LED-64 [197]	142	2048	19,008	1438	810
LED-64 [197] (Revised)	2176	2048	19,008	1438	1147

Table 4.6 Quantum resource requirements for LED and other lightweight block ciphers

Quantum circuit	#Qubits	#Toffoli	#CNOT	#X	Circuit depth
LED-64 (This work)	128	2048	8768	1158	753
LED-128 (This work)	192	3072	13,120	1734	1127
LED-64 (Corrected) [197]	2176	2048	19,008	1438	1147
PRESENT-64/80 [106]	144	2108	4683	1118	311
PRESENT-64/128 [106]	192	2232	4838	1164	311
GIFT-64/128 [106]	192	1792	1792	3261	308
GIFT-128/128 [106]	256	6144	6144	10,953	528
SIMON-64/128 [6]	192	1408	7396	1216	2643
SIMON-128/128 [6]	256	4352	17,152	4224	8427
SPECK-64/128 [103]	193	3286	9238	57	–
SPECK-128/128 [103]	257	7942	22,086	75	–
CHAM-64/128 [103]	196	2400	12,285	240	–
CHAM-128/128 [103]	268	4960	26,885	240	–

one MixColumnsSerial, 2048 qubits (32 rounds $\times$ 64) should be allocated for their implementation. However, their paper reports that only 142 qubits are required for the quantum LED-64 implementation without using any special techniques. This discrepancy arises from an issue in their MixColumnsSerial and naïve MixColumn implementation. To address this, we correctly allocate the output qubits for the naïve MixColumn and MixColumnsSerial in their implementation.

In Table 4.5, we report the corrected quantum resource requirements for their LED-64 quantum circuit. In Song et al.'s paper [197], all submodules of the quantum LED cipher are implemented without ancilla qubits, except for MixColumnsSerial. Consequently, the revised benchmark includes an additional 2048 qubits from the corrected MixColumnsSerial, bringing the total to 2176 qubits when combined with the initial 128 qubits allocated for plaintext (64 qubits) and key (64 qubits). Furthermore, the circuit depth is revised from 810 to 1147.

4.3.4.2 Experiment 1: Cost Analysis for Quantum LED

In this subsection, we present a detailed cost analysis of the proposed quantum implementations of LED cipher. Specifically, the comparison details between our designs and various implementations of quantum lightweight block ciphers from previous researches are provided in Table 4.6:

- **Our Work vs. Previous Quantum LED Design.** The proposed quantum LED-64 design in this work significantly outperforms the corrected LED-64 implementation discussed in Sect. 4.3.4.1, which represents the only previous quantum implementation of the LED cipher. Specifically, it requires fewer qubits, with a total of 128 compared to 2176 in the previous implementation, representing a reduction of approximately 94%. In terms of gate count, both designs use the same number of Toffoli gates, totaling 2048. However, our design shows clear advancements in other gate costs. Specifically, it utilizes fewer CNOT gates, with 8768 compared to 19,008 in the previous implementation, resulting in a reduction of approximately 53.9%. Similarly, our design requires fewer X gates, with 1158 compared to 1438, which corresponds to a reduction of around 19.5%. Regarding circuit depth, our design has a depth of 753, whereas the previous implementation has a depth of 1147. This represents a reduction of approximately 34.4%, highlighting a significant improvement in time efficiency for our proposed design.
- **Our Work vs. Other Quantum Analysis Lightweight Block Ciphers.** Our quantum LED designs demonstrate significant advantages in efficiency when compared to other quantum lightweight block cipher implementations. For qubits, our quantum LED-64 and LED-128 designs utilize 128 and 192 qubits, respectively. Compared to all the other lightweight block ciphers mentioned in this section, our designs achieve the lowest qubit requirements. Compared to other quantum block ciphers, our proposed quantum LED designs exhibit competitive gate efficiency. For the most resource-intensive quantum gate, the Toffoli gate, both LED-64 and LED-128 maintain a relatively low count, comparable to quantum PRESENT implementations [106]. Additionally, our designs show comparable efficiency in CNOT and X gates, requiring fewer resources than ciphers such as SIMON [6] and GIFT [106]. In terms of circuit depth, our designs exhibit a relatively high cost compared to other work, ranking just below the quantum Simon designs reported in [6]. While our designs maintain competitive performance, there is still room for optimization in circuit depth.

In brief, our LED-64 and LED-128 quantum implementations demonstrate the lowest qubit requirements when compared to previous quantum block cipher implementations, while also achieving competitive gate cost and circuit depth. This analysis clearly highlights LED as a strong candidate for lightweight block cipher applications.

4.3.4.3 Experiment 2: Grover's Attack on LED

This experiment focuses on estimating the quantum resources necessary for executing Grover's attack on the proposed LED ciphers. This process begins with a critical step—accurately estimating the quantum LED implementations. Specifically, the quantum LED-64 and LED-128 are implemented according to the methodology outlined in Sect. 4.3.3.1, followed by the decomposition of all involved Toffoli

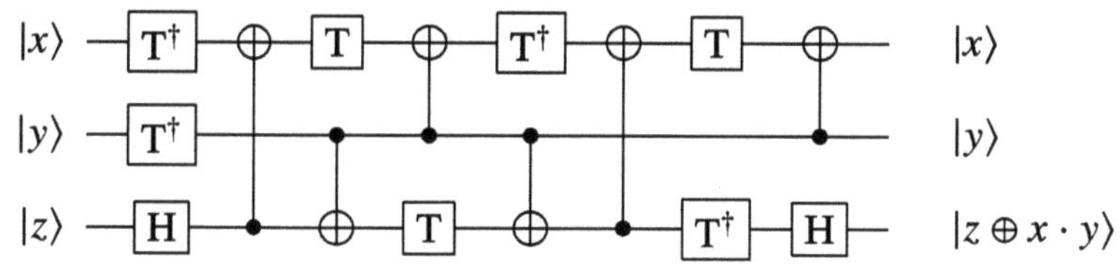

Fig. 4.14 Decomposition for Toffoli gate (T-depth 4, full depth 8)

gates into Clifford and T gates to ensure precise resource estimation. Among various decomposition methods available [5, 90, 189], this work adopts the approach introduced in [5]. As shown in Fig. 4.14, this method utilizes eight Clifford and seven T gates, resulting in a T-depth of 4 and a total depth of 8 for each Toffoli gate. Based on this specific decomposition strategy, we outlined the quantum resources required for the proposed LED implementations in Table 4.7.

In Tables 4.7 and 4.8, TD/Td-M and FD-M represent the trade-off performance of quantum circuits by being the product of the Toffoli/T depth and qubit count, and the full depth and qubit count, respectively. Additionally, we report the metrics TD^2/Td^2-M and FD^2-M, which are major trade-off metrics for parallelization in Grover's search,[8] emphasizing the importance of depth.

Based on the quantum cost of the proposed LED implementations, the corresponding Grover attack cost can be directly estimated. As detailed in Sect. 4.3.2.3, recovering a k-bit key for a cipher using Grover's algorithm requires approximately $\sqrt{2^k}$ iterations of both the Grover oracle and the diffusion operator. Moreover, [23] provides a precise analysis, indicating that the optimal number of iterations is $\lfloor \frac{\pi}{4}\sqrt{2^k} \rfloor$. In this chapter, we adopt this estimate for our cost calculations.

Furthermore, we employed a widely used method [72, 107] to simplify our analysis. Particularly, we disregard the diffusion operator from the cost estimation due to its minimal contribution. Consequently, the focus of Experiment 2 remains solely on the cost of the oracle, ensuring an accurate and efficient estimation process.

Next, we delve into the quantum resource analysis for Grover's attack oracle. This oracle consists of several components: the LED quantum circuit for encryption, an n-controlled NOT gate (where n represents the ciphertext size) for comparing the ciphertext with a known value, and the reverse operation of the previously executed LED quantum circuit for each subsequent iteration. It is crucial to emphasize the decomposition of the n-controlled NOT gate. Based on the approach outlined in [233], this gate is estimated to require $(32 \cdot n - 64)$ T gates.

[8] Grover's key search demands extreme circuit depth due to the large number of iterations. Under the constraint of circuit depth (such as the MAXDEPTH parameter introduced by NIST [151, 152]), parallelization of Grover's search is required to reduce the circuit depth, but its performance is poor. As a result, the product of the squared depth and qubit count serves as the trade-off metric in Grover's parallelization.

Table 4.7 Decomposed resource requirements for proposed quantum LED implementations

Circuit	#CNOT	#1qCliff	#T	Toffoli depth (TD)	#Qubit (M)	Full depth (FD)	TD-M	FD-M	TD^2-M	FD^2-M
LED-64	20,032	5254	14,336	128	128	1649	$1.00 \cdot 2^{14}$	$1.61 \cdot 2^{17}$	$1.00 \cdot 2^{21}$	$1.3 \cdot 2^{28}$
LED-128	30,016	7878	21,504	192	192	2471	$1.13 \cdot 2^{15}$	$1.81 \cdot 2^{18}$	$1.69 \cdot 2^{22}$	$1.09 \cdot 2^{30}$

Table 4.8 Quantum resource requirements for Grover's attack on LED

Cipher	#Gate (G)	Full depth (FD)	T-depth (Td)	#Qubit (M)	G-FD	FD-M	Td-M	FD^2-M	Td^2-M
LED-64	$1.99 \cdot 2^{47}$	$1.27 \cdot 2^{43}$	$1.57 \cdot 2^{41}$	$1.01 \cdot 2^{7}$	$1.26 \cdot 2^{91}$	$1.28 \cdot 2^{50}$	$1.58 \cdot 2^{48}$	$1.61 \cdot 2^{93}$	$1.24 \cdot 2^{90}$
LED-128	$1.47 \cdot 2^{81}$	$1.90 \cdot 2^{75}$	$1.18 \cdot 2^{74}$	$1.00 \cdot 2^{8}$	$1.39 \cdot 2^{157}$	$1.90 \cdot 2^{83}$	$1.18 \cdot 2^{82}$	$1.80 \cdot 2^{159}$	$1.39 \cdot 2^{156}$

Overall, the estimated costs for Grover's attack oracles on LED are summarized in Table 4.8. The total cost required is approximated as $\lfloor \frac{\pi}{4}\sqrt{2^k} \rfloor \times$ (Table 4.7 $\times$ 2) + $\lfloor \frac{\pi}{4}\sqrt{2^k} \rfloor \times (32 \cdot n - 64)$ T gates. Since the iterations are executed sequentially, the number of qubits required remains consistent with the values presented in Table 4.7, with only one additional decision qubit required for ciphertext comparison.

4.3.4.4 Experiment 3: STO Attack on Quantum LED

In this experiment, we present a detailed analysis of the STO attack on the proposed quantum LED design, with associated quantum costs outlined in Table 4.9:

- **STO Attack on Proposed Quantum LED-64.** The STO attack does not offer significant advantages over the Grover's oracle for attacking the LED-64 cipher. The parameter r is calculated as $r = \lceil k/n \rceil$, where k is the key size and n is the state size. A quantum attack on a block cipher requires at least r pairs of plaintext and corresponding ciphertext, where k denotes the key size and n denotes the block size. For LED-64, with both the block size and key size set to 64 bits, the required r is only 1. Hence, the Grover's attack oracle remains equally efficient when compared to the cheap oracle O_γ in STO attack. In brief, the STO attack does not provide a substantial advantage over Grover's attack oracle for the LED-64 cipher.
- **STO Attack on Proposed Quantum LED-128.**The STO attack offers significant improvement over Grover's attack for the LED-128 cipher. For LED-128, with a key size of 128 bits and a block size of 64 bits, the required number of rounds r is 2. Hence, the STO attack benefits from the cheap oracle O_γ, which is much more efficent compared to the corresponding Grover's oracle. While the STO attack requires a higher depth, it compensates by reducing the number of gates and qubits involved. This efficiency in terms of gate count and qubit cost highlights the advantages of the STO approach over Grover's attack for LED-128.

Overall, we have successfully implemented the STO attack on the proposed quantum LED ciphers, resulting in improved efficiency compared to Grover's attack. Although the STO attack does not offer significant advantages for LED-64, it is clear that for LED-128, STO attack oracle requires fewer gates and qubits, though with an increase in circuit depth. Moreover, the product of Gate and Full depth (G-FD) decreases, further highlighting the benefits of the STO approach in attacking LED-128.

Table 4.9 Quantum resource requirements for one oracle (Grover) and two oracles (STO) attacks on LED

Cipher	#Gate (G)	Full depth (FD)	T-depth (Td)	#Qubit (M)	G-FD	FD-M	Td-M	FD^2-M	Td^2-M
LED-64 (Grover)	$1.99 \cdot 2^{47}$	$1.27 \cdot 2^{43}$	$1.57 \cdot 2^{41}$	$1.01 \cdot 2^{7}$	$1.26 \cdot 2^{91}$	$1.28 \cdot 2^{50}$	$1.58 \cdot 2^{48}$	$1.61 \cdot 2^{93}$	$1.24 \cdot 2^{90}$
LED-64 (STO)	$1.99 \cdot 2^{47}$	$1.27 \cdot 2^{43}$	$1.57 \cdot 2^{41}$	$1.01 \cdot 2^{7}$	$1.26 \cdot 2^{91}$	$1.28 \cdot 2^{50}$	$1.58 \cdot 2^{48}$	$1.61 \cdot 2^{93}$	$1.24 \cdot 2^{90}$
LED-128 (Grover)	$1.47 \cdot 2^{81}$	$1.90 \cdot 2^{75}$	$1.18 \cdot 2^{74}$	$1.00 \cdot 2^{8}$	$1.39 \cdot 2^{157}$	$1.90 \cdot 2^{83}$	$1.18 \cdot 2^{82}$	$1.80 \cdot 2^{159}$	$1.39 \cdot 2^{156}$
LED-128 (STO)	$1.03 \cdot 2^{81}$	$1.33 \cdot 2^{76}$	$1.65 \cdot 2^{74}$	$1.51 \cdot 2^{7}$	$1.36 \cdot 2^{157}$	$1.99 \cdot 2^{83}$	$1.24 \cdot 2^{82}$	$1.32 \cdot 2^{160}$	$1.02 \cdot 2^{157}$

Table 4.10 Quantum resource requirements for LED under various modes of operation with η blocks

Quantum circuit	#Qubits	#Toffoli	#CNOT	#X	Circuit depth
LED-64 (ECB)	$64 + 64\eta$	2048η	8768η	1158η	753η
LED-128 (ECB)	$128 + 64\eta$	3072η	$13,120\eta$	1734η	1127η
LED-64 (CBC)	$128 + 64\eta$	2048η	8832η	1158η	754η
LED-128 (CBC)	$192 + 64\eta$	3072η	$13,184\eta$	1734η	1128η
LED-64 (CFB)	$64 + 128\eta$	2048η	$8896\eta - 64$	1158η	$755\eta - 1$
LED-128 (CFB)	$128 + 128\eta$	3072η	$13,248\eta - 64$	1734η	$1129\eta - 1$

4.3.4.5 Experiment 4: Grover's Attack on LED Across Various Modes of Operation

In the last experiment, we perform Grover's attack on the proposed quantum LED cipher across multiple modes of operation, including ECB, CBC, and CFB.

Before estimating the resources required for Grover's attack, we first analyze the quantum costs associated with the different modes of operation frameworks discussed in Sect. 4.3.3.2. According to these frameworks, Table 4.10 provides the detailed quantum resource requirements for η-block quantum LED encryption across different modes of operation, showing how the cost of each quantum framework varies with block size η.

However, regardless of the mode of operation used, Grover's attack only needs to be applied to the first block of the cipher. This is because all blocks within a specific mode generally share the same key. Consequently, as shown in Table 4.11, the cost estimation for the quantum attack exhibits only minor differences across different modes of operation.

In brief, the cost of Grover's attack depends solely on the first block, and the quantum resource requirements for this block vary only slightly between ECB, CBC, and CFB. Therefore, there is minimal variation in the overall cost of Grover's attack across different modes of operation.

4.3.5 Section Summary

In this chapter, we have designed efficient quantum implementation of the LED block cipher. By optimizing the quantum circuits through minimization of qubits, quantum gates, and circuit depth, we were able to estimate the least quantum resources required for executing Grover's attack and STO attack against LED. Furthermore, we have successfully implemented ECB, CBC, and CFB modes of operation for quantum LED; this kind of analysis was not done in the past, to best of our knowledge.

Table 4.11 Quantum resource requirements for Grover's attack on LED across various modes of operations

Cipher	#Gate (G)	Full depth (FD)	T-depth (Td)	#Qubit (M)	G-FD	FD-M	Td-M	FD^2-M	Td^2-M
LED-64 (ECB)	$1.99 \cdot 2^{47}$	$1.27 \cdot 2^{43}$	$1.57 \cdot 2^{41}$	$1.01 \cdot 2^{7}$	$1.26 \cdot 2^{91}$	$1.28 \cdot 2^{50}$	$1.58 \cdot 2^{48}$	$1.61 \cdot 2^{93}$	$1.24 \cdot 2^{90}$
LED-128 (ECB)	$1.47 \cdot 2^{81}$	$1.90 \cdot 2^{75}$	$1.18 \cdot 2^{74}$	$1.00 \cdot 2^{8}$	$1.39 \cdot 2^{157}$	$1.90 \cdot 2^{83}$	$1.18 \cdot 2^{82}$	$1.80 \cdot 2^{159}$	$1.39 \cdot 2^{156}$
LED-64 (CBC)	$1.99 \cdot 2^{47}$	$1.27 \cdot 2^{43}$	$1.57 \cdot 2^{41}$	$1.51 \cdot 2^{7}$	$1.26 \cdot 2^{91}$	$1.91 \cdot 2^{50}$	$1.18 \cdot 2^{49}$	$1.20 \cdot 2^{94}$	$1.85 \cdot 2^{90}$
LED-128 (CBC)	$1.47 \cdot 2^{81}$	$1.90 \cdot 2^{75}$	$1.18 \cdot 2^{74}$	$1.25 \cdot 2^{8}$	$1.39 \cdot 2^{157}$	$1.19 \cdot 2^{84}$	$1.475 \cdot 2^{82}$	$1.13 \cdot 2^{160}$	$1.74 \cdot 2^{156}$
LED-64 (CFB)	$1.99 \cdot 2^{47}$	$1.27 \cdot 2^{43}$	$1.57 \cdot 2^{41}$	$1.51 \cdot 2^{7}$	$1.26 \cdot 2^{91}$	$1.91 \cdot 2^{50}$	$1.18 \cdot 2^{49}$	$1.20 \cdot 2^{94}$	$1.85 \cdot 2^{90}$
LED-128 (CFB)	$1.47 \cdot 2^{81}$	$1.90 \cdot 2^{75}$	$1.18 \cdot 2^{74}$	$1.25 \cdot 2^{8}$	$1.39 \cdot 2^{157}$	$1.19 \cdot 2^{84}$	$1.475 \cdot 2^{82}$	$1.13 \cdot 2^{160}$	$1.74 \cdot 2^{156}$

4.4 Chapter Summary

In this chapter, we explore two impactful applications of quantum arithmetic: quantum machine learning and quantum cryptanalysis, each showcasing quantum arithmetic's potential to enhance computational efficiency across various domains.

Firstly, we investigate the mapping of our design onto real quantum hardware architectures, with a particular emphasis on verifying our designs using available quantum hardware. Our primary focus is on NISQ devices, which are currently the only accessible hardware architecture in quantum computing.

In the first application, quantum machine learning, we conducted the work "Efficient Quantum Circuits for Machine Learning Activation Functions with Constant T-Depth ReLU." My contribution focused on designing an efficient quantum leaky ReLU activation function, achieving a constant T-depth of 8. This work supports the integration of activation functions into fault-tolerant quantum computing architectures, enhancing the practicality of quantum machine learning.

The second application explored quantum cryptanalysis domain through our work, "New Results in Quantum Analysis of LED: Featuring One and Two Oracle Attacks." This work examined the lightweight block cipher LED, which is essential for resource-constrained environments such as the Internet of Things. We proposed an optimized quantum analysis circuit for LED that minimizes overall quantum resource requirements. Specifically, we performed Grover's attack and the search with two Oracles (STO) attack on the LED cipher, while also evaluating the efficiency and effectiveness of the quantum attack across different modes of operation.

Key Takeaways

1. This chapter explores two significant applications of quantum arithmetic: quantum machine learning and quantum cryptanalysis.
2. We focus on mapping and verifying designs on NISQ devices, which are currently the most accessible quantum hardware.
3. In quantum machine learning, we introduced an efficient quantum leaky ReLU activation function with a constant T-depth of 8, supporting fault-tolerant quantum computing architectures.
4. In quantum cryptanalysis, we optimized the quantum analysis circuit for the LED cipher, minimizing quantum resource requirements, and evaluated the effectiveness of Grover's and search with two Oracles (STO) attacks.

Chapter 5
Conclusion and Future Roadmap

By 2100, our destiny is to become like the gods we once worshipped and feared. But our tools will not be magic wands and potions but the science of computers, nanotechnology, artificial intelligence, biotechnology, and most of all, the quantum theory.

–Michio Kaku

5.1 Conclusion

The exploration of quantum arithmetic circuits is not only interesting but also challenging. In our current research stage, the primary aim of this book is to enhance the performance of quantum arithmetic circuits by integrating classical design techniques, bringing improved efficiency and scalability to quantum computing. Additionally, it includes mapping these designs onto real quantum hardware and exploring potential applications across diverse fields. Over the past few decades, quantum computing has witnessed substantial progress, providing a foundational platform for our present research. Here, we sincerely acknowledge the invaluable contributions of the previous researchers in this field, who provided us with so many important insights and inspiration for our ongoing research.

In this book, the main achievements are outlined as follows:[1]

- **Achievement 1: Optimization of Quantum Arithmetic Designs**
  This book presents significant advancements in the optimization of quantum arithmetic designs, focusing on enhancing quantum efficiency and minimizing overall time. By employing innovative methodologies such as higher radix strategies, Ling structures, and Slansky architectures, this work introduces several optimized frameworks for quantum addition [224–226, 232], modular addition [231], and division operations [230]. Our researches not only improve

[1] Apart from the work presented in this book, I have also contributed as a co-author to Papers [10, 126, 169, 225, 235].

S. Wang, A. Chattopadhyay, *Quantum Arithmetic*, Computer Architecture and Design Methodologies, https://doi.org/10.1007/978-981-95-6039-4_5

the performance of quantum arithmetic circuits but also lay the groundwork for future exploration in the field of quantum arithmetic.

- **Achievement 2: Comprehensive Overview of Quantum Arithmetic**
  In this book, we offer a thorough analysis of the latest innovations in quantum arithmetic [229], focusing on fundamental operations such as addition, subtraction, multiplication, division, and modular exponentiation. By illuminating key developments in quantum arithmetic domain, the work not only highlights significant advancements but also identifies existing gaps within the literature that warrant further investigation. Therefore, it serves as a crucial reference for researchers, fostering further exploration and encouraging the evolution in the quantum computing landscape.
- **Achievement 3: Contributions to Quantum Cybersecurity**
  Apart from NISQ devices, we also investigate the ion-trapped quantum devices. Quantum computers hold the potential to revolutionize various fields, including financial portfolio optimization, drug discovery, and cryptography, by providing algorithmic speedups over classical methods. However, their widespread adoption also makes them attractive targets for cybercriminals. Despite this risk, there has been limited experimental research on the potential attacks against real quantum systems. In this chapter, we introduce MalaQ [228], a malware specifically designed to exploit vulnerabilities in the classical front end of ion-trapped quantum computers, resulting in performance degradation and potential circuit failure. By highlighting these vulnerabilities, our research provides essential insights into quantum cybersecurity.
- **Achievement 4: Exploration of Quantum Applications**
  This book explores the applications of quantum arithmetic circuits, with a particular focus on quantum machine learning (QML) [252] and quantum cryptanalysis [227]. By examining these applications, our research highlights the practicality of quantum arithmetic across various domains.

These researches demonstrate a fruitful research journey, ranging from efficient quantum arithmetic designs to their relevant applications. Our exploration not only introduces novel designs but also extends into its broader implications for quantum machine learning and quantum cryptanalysis, resulting in comprehensive and in-depth investigations. Obviously, each achievement of this book provides unique insights and perspectives within the field of quantum arithmetic.

In conclusion, this work collectively advances the efficiency of quantum arithmetic circuits, establishing a robust foundation for future exploration. Our efforts contribute to the realization of efficient and scalable quantum arithmetic designs, offering innovative solutions to significant challenges in quantum computing and other domains. Looking ahead, further research could focus on other quantum arithmetic designs, such as quantum multiplication and quantum exponentiation, while also investigating relevant applications in fields like quantum cryptanalysis and beyond.

5.2 Future Roadmap

This section provides a detailed description of my future research plans. After completing my Ph.D. research, I intend to shift the research focus from quantum arithmetic circuits to large-scale quantum algorithms. This transition aims to facilitate a comprehensive exploration of the quantum domain from bottom to top.

5.2.1 Exploration of Quantum Arithmetic

- **Approximate Division in Quantum Arithmetic Circuits**
  Inspired by the Newton-Raphson approximate division in the classical domain, we explore new techniques for approximate division in quantum computing, addressing challenges and opportunities specific to quantum division.
- **Quantum Multiplier Design**
  Further advance quantum multiplier design by incorporating insights gained from the optimization of quantum carry-save modular adders and quantum dividers. Explore novel approaches to enhance the efficiency and speed of quantum multiplication.
- **Compressor-Based Quantum Arithmetic**
  In classical arithmetic, compressor is a more generalized approach to constructing arithmetic circuits compared to traditional half-adder and full-adder-based methods. Therefore, this project aims to propose a range of quantum arithmetic designs using compressors to explore the benefits of more generalized representations. We will evaluate potential improvements in overall efficiency compared to existing designs and investigate whether compressors can provide further optimization opportunities within quantum arithmetic.

5.2.2 Transition to Large-Scale Algorithms

- **Error correction of quantum arithmetic circuits and mapping on real quantum computers**
  Investigating error correction mechanisms for prominent quantum arithmetic circuits and the corresponding costs on real quantum computers. In this chapter, we will conduct experiments and comparisons of existing quantum arithmetic circuits on real quantum hardware to explore their real efficiency. The exploration aims to assess the consistency between theory and experimentation.
- **Quantum Adder Trade-offs in SHA-2 Analysis**
  Conduct a quantum analysis of SHA-2 to explore the trade-offs introduced by different quantum addition blocks within the quantum framework. This study will examine how various adder designs impact the efficiency of SHA-2 quantum

analysis, providing insights into optimal configurations for the corresponding quantum implementations.

5.2.3 Exploration of Large-Scale Algorithms

- **Framework of Shor's Algorithm**
  We will develop a quantum framework for Shor's algorithm, which can be utilized to perform efficiency analyses on different quantum arithmetic blocks. Besides, this framework will enable us to conduct quantum analyses on various ciphers.
- **Quantum Machine Learning**
  We will also delve into various directions of quantum machine learning, such as classification. In these researches, our target is to explore the potential applications of quantum machine learning in diverse contexts and propose innovative quantum solutions to propel the development of machine learning.
- **Quantum Crosstalk Attack**
  We will investigate crosstalk effects within the IBM quantum framework, as this topic is critical for quantum cybersecurity. Moreover, we will explore crosstalk attacks to extract information from the victim's quantum circuits.

References

1. Abbas, A., Sutter, D., Zoufal, C., Lucchi, A., Figalli, A., Woerner, S.: The power of quantum neural networks. Nat. Comput. Sci. **1**(6), 403–409 (2021)
2. Acharya, R., Abanin, D.A., Aghababaie-Beni, L., Aleiner, I., Andersen, T.I., Ansmann, M., Arute, F., Arya, K., Asfaw, A., Astrakhantsev, N., et al.: Quantum error correction below the surface code threshold. Nature **638**, 920–926 (2024)
3. Affonso, C., Rossi, A.L.D., Vieira, F.H.A., de Leon Ferreira, A.C.P., et al.: Deep learning for biological image classification. Expert Syst. Appl. **85**, 114–122 (2017)
4. AI, G.Q.: Suppressing quantum errors by scaling a surface code logical qubit. Nature **614**(7949), 676–681 (2023)
5. Amy, M., Maslov, D., Mosca, M., Roetteler, M., Roetteler, M.: A meet-in-the-middle algorithm for fast synthesis of depth-optimal quantum circuits. IEEE Trans. Comput. Aided Des. Integr. Circuits Syst. **32**(6), 818–830 (2013). https://doi.org/10.1109/TCAD.2013.2244643
6. Anand, R., Maitra, A., Mukhopadhyay, S.: Grover on simon. Quantum Inf. Process. **19**(9), Article number 340 (2020). https://doi.org/10.1007/s11128-020-02844-w
7. AnanthaLakshmi, A., Sudha, G.F.: A novel power efficient 0.64-gflops fused 32-bit reversible floating point arithmetic unit architecture for digital signal processing applications. Microprocess. Microsyst. **51**, 366–385 (2017)
8. Asadi, M.A., Mosleh, M., Haghparast, M.: Towards designing quantum reversible ternary multipliers. Quantum Inf. Process. **20**(7), 226 (2021)
9. Atchade-Adelomou, P., Gonzalez, S.: Efficient quantum modular arithmetics for the isq era (2023). https://doi.org/10.48550/ARXIV.2311.08555. https://arxiv.org/abs/2311.08555
10. Baksi, A., Chakraborty, S., Chattopadhyay, A., Chun, M., Islam, S.H., Jang, K., Kim, H., Oh, Y., Roy, S., Seo, H., Wang, S.: Quantum implementation of linear and non-linear layers. In: IEEE International System-on-Chip Conference (SOCC) (2024)
11. Baksi, A., Karmakar, B., Dasu, V.A.: POSTER: optimizing device implementation of linear layers with automated tools. In: Applied Cryptography and Network Security Workshops - ACNS 2021 Satellite Workshops, Kamakura, Japan, June 21–24, 2021, Proceedings. Lecture Notes in Computer Science, vol. 12809, pp. 500–504. Springer (2021). https://doi.org/10.1007/978-3-030-81645-2_30
12. Barak, B., Marwaha, K.: Classical algorithms and quantum limitations for maximum cut on high-girth graphs. arXiv preprint arXiv:2106.05900 (2021)
13. Beauregard, S.: Circuit for shor's algorithm using 2n+3 qubits (2002). https://doi.org/10.48550/ARXIV.QUANT-PH/0205095. https://arxiv.org/abs/quant-ph/0205095
14. Beckman, D., Chari, A.N., Devabhaktuni, S., Preskill, J.: Efficient networks for quantum factoring. Phys. Rev. A **54**(2), 1034–1063 (1996). https://doi.org/10.1103/PhysRevA.54.1034

S. Wang, A. Chattopadhyay, *Quantum Arithmetic*, Computer Architecture and Design Methodologies, https://doi.org/10.1007/978-981-95-6039-4

15. Benedetti, M., Lloyd, E., Sack, S., Fiorentini, M.: Parameterized quantum circuits as machine learning models. Quantum Sci. Technol. **4**(4), 043001 (2019)
16. Bennett, C.H.: Logical reversibility of computation. Ibm J. Res. Dev. **17**, 525–532 (1973). https://api.semanticscholar.org/CorpusID:14641793
17. Berg, E., Minev, Z., Kandala, A., Temme, K.: Probabilistic error cancellation with sparse pauli–lindblad models on noisy quantum processors. Nat. Phys. **19**, 1–6 (2023). https://doi.org/10.1038/s41567-023-02042-2
18. Biamonte, J., Wittek, P., Pancotti, N., Rebentrost, P., Wiebe, N., Lloyd, S.: Quantum machine learning. Nature **549**(7671), 195–202 (2017). https://doi.org/10.1038/nature23474
19. Black, F., Scholes, M.: The pricing of options and corporate liabilities. J. Polit. Econ. **81**(3), 637–654 (1973). https://doi.org/10.1086/260062
20. Bluvstein, D., Evered, S.J., Geim, A.A., Li, S.H., Zhou, H., Manovitz, T., Ebadi, S., Cain, M., Kalinowski, M., Hangleiter, D., et al.: Logical quantum processor based on reconfigurable atom arrays. Nature **626**(7997), 58–65 (2024)
21. Bogdanov, A., Knudsen, L.R., Leander, G., Paar, C., Poschmann, A., Robshaw, M.J., Seurin, Y., Vikkelsoe, C.: PRESENT: An ultra-lightweight block cipher. In: CHES, vol. 4727, pp. 450–466. Springer (2007)
22. Boole, G.: The mathematical analysis of logic. Philosophical Library (1847)
23. Boyer, M., Brassard, G., Høyer, P., Tapp, A.: Tight bounds on quantum searching. Fortschritte der Physik **46**(4-5), 493–505 (1998). https://doi.org/10.1002/(SICI)1521-3978(199806)46:4/5<493::AID-PROP493>3.0.CO;2-P
24. Bravyi, S., Cross, A.W., Gambetta, J.M., Maslov, D., Rall, P., Yoder, T.J.: High-threshold and low-overhead fault-tolerant quantum memory. Nature **627**(8005), 778–782 (2024)
25. Brent, R., Kung, H.: A regular layout for parallel adders. IEEE Trans. Comput. **31**, 260–264 (2004). https://doi.org/10.1109/TC.1982.1675982
26. Cai, B., Gao, F., Leander, G.: Quantum attacks on two-round even-mansour. Front. Phys. **10**, 1028014 (2022)
27. Camps, D., Lin, L., Van Beeumen, R., Yang, C.: Explicit quantum circuits for block encodings of certain sparse matrices (2022). https://doi.org/10.48550/ARXIV.2203.10236. https://arxiv.org/abs/2203.10236
28. Cerezo, M., Verdon, G., Huang, H.Y., Cincio, L., Coles, P.J.: Challenges and opportunities in quantum machine learning. Nat. Comput. Sci. **2**(9), 567–576 (2022). https://doi.org/10.1038/s43588-022-00311-3
29. Chakrabarti, S., Krishnakumar, R., Mazzola, G., Stamatopoulos, N., Woerner, S., Zeng, W.J.: A threshold for quantum advantage in derivative pricing. Quantum **5**, 463 (2021). https://doi.org/10.22331/q-2021-06-01-463
30. Chen, J., Liu, Q., Fan, Y., Wu, L., Li, B., Wang, M.: New SAT-based model for quantum circuit decision problem: Searching for low-cost quantum implementation. IACR Commun. Cryptol. **1**(1), (2024). https://doi.org/10.62056/anmmp-4c2h
31. Chen, S., Xu, H., Liu, D., Hu, B., Wang, H.: A vision of iot: Applications, challenges, and opportunities with china perspective. IEEE Internet Things J. **1**(4), 349–359 (2014)
32. Cheng, K.W., Tseng, C.C.: Quantum full adder and subtractor. Electron. Lett. **38**(22), 1343–1344 (2002)
33. Chun, M., Baksi, A., Chattopadhyay, A.: Dorcis: Depth optimized quantum implementation of substitution boxes. Cryptology ePrint Archive, Paper 2023/286 (2023). https://eprint.iacr.org/2023/286
34. Clader, B.D., Dalzell, A.M., Stamatopoulos, N., Salton, G., Berta, M., Zeng, W.J.: Quantum resources required to block-encode a matrix of classical data. IEEE Trans. Quantum Eng. **3**, 1–23 (2022). https://doi.org/10.1109/TQE.2022.3231194
35. Cleve, R., Watrous, J.: Fast parallel circuits for the quantum fourier transform. In: Proceedings 41st Annual Symposium on Foundations of Computer Science, pp. 526–536. IEEE (2000)
36. Cong, I., Choi, S., Lukin, M.D.: Quantum convolutional neural networks. Nat. Phys. **15**(12), 1273–1278 (2019)

37. Cook, S.A., Aanderaa, S.O.: On the minimum computation time of functions. Trans. Am. Math. Soc. **142**, 291–314 (1969)
38. Courtois, N.T., Hulme, D., Mourouzis, T.: Solving circuit optimisation problems in cryptography and cryptanalysis. Cryptology ePrint Archive, Paper 2011/475 (2011). https://eprint.iacr.org/2011/475
39. Cross, A., Bishop, L., Sheldon, S., Nation, P., Gambetta, J.: Validating quantum computers using randomized model circuits. Phys. Rev. A **100**, 032328 (2019). https://doi.org/10.1103/PhysRevA.100.032328
40. Cuccaro, S., Draper, T., Kutin, S., Moulton, D.: A new quantum ripple-carry addition circuit. arXiv (2008). https://arxiv.org/pdf/quant-ph/0410184.pdf
41. Das, R., Chattopadhyay, A., Rahaman, H.: Optimizing quantum circuits for modular exponentiation. In: 2019 32nd International Conference on VLSI Design and 2019 18th International Conference on Embedded Systems (VLSID), pp. 407–412. IEEE (2019)
42. Dasu, V.A., Baksi, A., Sarkar, S., Chattopadhyay, A.: Lighter-r: Optimized reversible circuit implementation for sboxes. In: 2019 32nd IEEE International System-on-Chip Conference (SOCC), pp. 260–265 (2019). https://api.semanticscholar.org/CorpusID:218564036
43. Davenport, J.H., Pring, B.: Improvements to quantum search techniques for block-ciphers, with applications to AES. In: International Conference on Selected Areas in Cryptography, pp. 360–384. Springer (2020)
44. Draper, T., Kutin, S., Rains, E., Svore, K.: A logarithmic-depth quantum carry-lookahead adder. Quantum Inf. Comput. **6**, 351–369 (2004). https://doi.org/10.26421/QIC6.4-5-4
45. Draper, T.G.: Addition on a quantum computer (2000). https://doi.org/10.48550/ARXIV.QUANT-PH/0008033. https://arxiv.org/abs/quant-ph/0008033
46. Draper, T.G., Kutin, S.A., Rains, E.M., Svore, K.M.: A logarithmic-depth quantum carry-lookahead adder (2004). https://doi.org/10.5555/2012086.2012090
47. Dutta, S., Bhattacharjee, D., Chattopadhyay, A.: Quantum circuits for toom-cook multiplication. Phys. Rev. A **98**, 012311 (2018). https://doi.org/10.1103/PhysRevA.98.012311. https://link.aps.org/doi/10.1103/PhysRevA.98.012311
48. Earle, J.: Latched carry-save adder. IBM Tech. Disclosure Bull. **7**(10), 909–910 (1965)
49. Earle, J.G.: Latched carry save adder circuit for multipliers (Sep 5 1967), uS Patent 3,340,388
50. Egger, D.J., Gambella, C., Marecek, J., McFaddin, S., Mevissen, M., Raymond, R., Simonetto, A., Woerner, S., Yndurain, E.: Quantum computing for finance: State-of-the-art and future prospects. IEEE Trans. Quantum Eng. **1**, 1–24 (2020). https://doi.org/10.1109/TQE.2020.3030314
51. Faghih, E., Taheri, M., Navi, K., Bagherzadeh, N.: Efficient realization of quantum balanced ternary reversible multiplier building blocks: A great step towards sustainable computing. Sustain. Comput. Inf. Syst. **40**, 100908 (2023)
52. Feng, J., Wei, Y., Zhang, F., Pasalic, E., Zhou, Y.: Novel optimized implementations of lightweight cryptographic s-boxes via sat solvers. IEEE Trans. Circuits Syst. I Regular Papers **71**(1), 334–347 (2024). https://doi.org/10.1109/TCSI.2023.3325559
53. Feynman, R.P.: Simulating physics with computers. Int. J. Theor. Phys. **21**(6/7), 467–488 (1982)
54. Fowler, A.G., Devitt, S.J., Hollenberg, L.C.L.: Implementation of shor's algorithm on a linear nearest neighbour qubit array (2004). https://doi.org/10.48550/ARXIV.QUANT-PH/0402196. https://arxiv.org/abs/quant-ph/0402196
55. Fowler, A.G., Mariantoni, M., Martinis, J.M., Cleland, A.N.: Surface codes: Towards practical large-scale quantum computation. Phys. Rev. A Atom. Mol. Opt. Phys. **86**(3), 032324 (2012)
56. Fu, X., Riesebos, L., Lao, L., Almudever, C.G., Sebastiano, F., Versluis, R., Charbon, E., Bertels, K.: A heterogeneous quantum computer architecture. In: Proceedings of the ACM International Conference on Computing Frontiers. p. 323–330. CF '16. Association for Computing Machinery, New York, NY, USA (2016). https://doi.org/10.1145/2903150.2906827
57. Fu, X., Bao, W., Zhou, C.: Speeding up implementation for shor's factorization quantum algorithm. Chin. Sci. Bull. **55**, 3648–3653 (2010)

58. Gayathri, S.S., Kumar, R., Dhanalakshmi, S.: Efficient floating-point division quantum circuit using newton-raphson division. J. Phys. Confer. Ser. **2335**(1), 012058 (2022). https://doi.org/10.1088/1742-6596/2335/1/012058
59. Gayathri, S.S., Kumar, R., Dhanalakshmi, S., Dooly, G., Duraibabu, D.B.: T-count optimized quantum circuit designs for single-precision floating-point division. Electronics **10**, 703 (2021). https://doi.org/10.3390/electronics10060703
60. Gayathri, S., Kumar, R., Dhanalakshmi, S.: Efficient floating-point division quantum circuit using newton-raphson division. J. Phys. Confer. Ser., **2335**, 012058 (2022). IOP Publishing
61. Gayathri, S., Kumar, R., Dhanalakshmi, S., Dooly, G., Duraibabu, D.B.: T-count optimized quantum circuit designs for single-precision floating-point division. Electronics **10**(6), 703 (2021)
62. Gidney, C.: Halving the cost of quantum addition. Quantum **2**, 74 (2018). https://doi.org/10.22331/q-2018-06-18-74
63. Gidney, C.: Halving the cost of quantum addition. Quantum **2**, 74 (2018). https://doi.org/10.22331/q-2018-06-18-74
64. Gidney, C.: Asymptotically efficient quantum karatsuba multiplication. arXiv preprint arXiv:1904.07356 (2019)
65. Gidney, C.: Windowed quantum arithmetic. arXiv preprint arXiv:1905.07682 (2019)
66. Gilyén, A., Su, Y., Low, G.H., Wiebe, N.: Quantum singular value transformation and beyond: exponential improvements for quantum matrix arithmetics. In: Proceedings of the 51st Annual ACM SIGACT Symposium on Theory of Computing, STOC '19. ACM (Jun 2019). https://doi.org/10.1145/3313276.3316366
67. Goel, S., Klivans, A.R.: Learning neural networks with two nonlinear layers in polynomial time. In: Conference on Learning Theory, pp. 1470–1499. PMLR (2019)
68. Gokhale, P., Koretsky, S., Huang, S., Majumder, S., Drucker, A., Brown, K.R., Chong, F.T.: Quantum fan-out: Circuit optimizations and technology modeling. In: 2021 IEEE International Conference on Quantum Computing and Engineering (QCE), pp. 276–290. IEEE (2021)
69. Goldschmidt, R.E.: Applications of division by convergence. Ph.D. thesis, Massachusetts Institute of Technology (1964)
70. Gossett, P.: Quantum carry-save arithmetic. arXiv preprint quant-ph/9808061 (1998)
71. Gottesman, D.: An introduction to quantum error correction and fault-tolerant quantum computation. In: Quantum Information Science and Its Contributions to Mathematics, Proceedings of Symposia in Applied Mathematics, vol. 68, pp. 13–58 (2010)
72. Grassl, M., Langenberg, B., Roetteler, M., Steinwandt, R.: Applying Grover's algorithm to AES: Quantum resource estimates. In: Takagi, T. (ed.) Post-Quantum Cryptography, pp. 29–43. Springer International Publishing, Cham (2016)
73. Griffiths, R.B., Niu, C.S.: Semiclassical fourier transform for quantum computation. Phys. Rev. Lett. **76**(17), 3228–3231 (1996). https://doi.org/10.1103/PhysRevLett.76.3228
74. Grover, L.K.: A fast quantum mechanical algorithm for database search. In: Proceedings of the Twenty-Eighth Annual ACM Symposium on Theory of Computing, pp. 212–219 (1996)
75. Guo, J., Peyrin, T., Poschmann, A., Robshaw, M.: The led block cipher. In: Cryptographic Hardware and Embedded Systems–CHES 2011: 13th International Workshop, Nara, Japan, September 28–October 1, 2011. Proceedings 13, pp. 326–341. Springer (2011)
76. Guo, N., Mitarai, K., Fujii, K.: Nonlinear transformation of complex amplitudes via quantum singular value transformation (2021). https://doi.org/10.48550/ARXIV.2107.10764. https://arxiv.org/abs/2107.10764
77. Gupta, S., Zia, R.: Quantum neural networks. J. Comput. Syst. Sci. **63**(3), 355–383 (2001). https://doi.org/10.1006/jcss.2001.1769
78. Gurney, K.: An Introduction to Neural Networks. CRC Press (2018)
79. Guseynov, N., Huang, X., Liu, N.: Explicit gate construction of block-encoding for hamiltonians needed for simulating partial differential equations (2024). https://doi.org/10.48550/ARXIV.2405.12855. https://arxiv.org/abs/2405.12855

80. Gükaynak, F., Leblebicit, Y., Chaouati, L., McGuinness, P.: Higher radix kogge-stone parallel prefix adder architectures. In: 2000 IEEE International Symposium on Circuits and Systems (ISCAS), vol. 5, pp. 609–612 (2000). https://doi.org/10.1109/ISCAS.2000.857516
81. Hallgren, S.: Polynomial-time quantum algorithms for pell's equation and the principal ideal problem. J. ACM **54**(1), 1–19 (2007). https://doi.org/10.1145/1206035.1206039
82. Han, T., Carlson, D.A.: Fast area-efficient vlsi adders. In: Proceedings - Symposium on Computer Arithmetic (1987). https://doi.org/10.1109/ARITH.1987.6158699
83. Häner, T., Roetteler, M., Svore, K.M.: Factoring using 2n + 2 qubits with toffoli based modular multiplication. Quantum Info. Comput. **17**(7–8), 673–684 (2017)
84. Häner, T., Roetteler, M., Svore, K.M.: Optimizing quantum circuits for arithmetic. arXiv preprint arXiv:1805.12445 (2018)
85. Hansen, E., Joshi, S., Rarick, H.: Resource estimation of quantum multiplication algorithms. In: 2023 IEEE International Conference on Quantum Computing and Engineering (QCE), vol. 2, pp. 199–202. IEEE (2023)
86. Harris, D.: A taxonomy of parallel prefix networks. In: The Thrity-Seventh Asilomar Conference on Signals, Systems & Computers, 2003, vol. 2, pp. 2213–2217 (2003). https://doi.org/10.1109/ACSSC.2003.1292373
87. Harris, D., Oberman, S., Horowitz, M.: Srt division: Architectures, models, and implementations. Computer Systems Library, Standard University, Tech. Rep (1998)
88. Harrow, A.W., Hassidim, A., Lloyd, S.: Quantum algorithm for linear systems of equations. Phys. Rev. Lett. **103**(15), 150,502 (2009). https://doi.org/10.1103/PhysRevLett.103.150502
89. Havlíček, V., Córcoles, A.D., Temme, K., Harrow, A.W., Kandala, A., Chow, J.M., Gambetta, J.M.: Supervised learning with quantum-enhanced feature spaces. Nature **567**(7747), 209–212 (2019). https://doi.org/10.1038/s41586-019-0980-2
90. He, Y., Luo, M.X., Zhang, E., Wang, H.K., Wang, X.F.: Decompositions of n-qubit toffoli gates with linear circuit complexity. Int. J. Theor. Phys. **56**(7), 2350–2361 (2017)
91. He, Y., Zhao, C., Dai, G., He, K., Geng, X., Liu, J., Chen, W.: Quantum modular multiplier via binary-exponent-based recombination. Quantum Inf. Process. **21**(12), 391 (2022)
92. Held, S., Spirkl, S.T.: Binary adder circuits of asymptotically minimum depth, linear size, and fan-out two. ACM Trans. Algorithms **14**(1), 1–18 (2017). https://doi.org/10.1145/3147215
93. Hendrycks, D., Gimpel, K.: Gaussian error linear units (gelus). arXiv preprint arXiv:1606.08415 (2016)
94. Herman, D., Googin, C., Liu, X., Sun, Y., Galda, A., Safro, I., Pistoia, M., Alexeev, Y.: Quantum computing for finance. Nat. Rev. Phys. **5**(8), 450–465 (2023). https://doi.org/10.1038/s42254-023-00603-1
95. Hoefler, T., Häner, T., Troyer, M.: Disentangling hype from practicality: On realistically achieving quantum advantage. Commun. ACM **66**, 82–87 (2023). https://doi.org/10.1145/3571725
96. Hoffman, R., Hayes, P., Ford, K.: Human-centered computing: Thinking in and out of the box. Intell. Syst. IEEE **16**, 76–78 (2001). https://doi.org/10.1109/5254.956085
97. Huang, H.Y., Bharti, K., Rebentrost, P.: Near-term quantum algorithms for linear systems of equations with regression loss functions. New J. Phys. **23**(11), 113021 (2021). https://doi.org/10.1088/1367-2630/ac325f
98. Huang, H.Y., Broughton, M., Mohseni, M., Babbush, R., Boixo, S., Neven, H., McClean, J.R.: Power of data in quantum machine learning. Nat. Commun. **12**(1), Article number: 2631 (2021). https://doi.org/10.1038/s41467-021-22539-9
99. Huang, P.W., Li, X., Koor, K., Rebentrost, P.: Hybrid quantum-classical and quantum-inspired classical algorithms for solving banded circulant linear systems (2023). https://doi.org/10.48550/ARXIV.2309.11451. https://arxiv.org/abs/2309.11451
100. Haener, T., Soeken, M., Roetteler, M., Svore, K.M.: Quantum circuits for floating-point arithmetic. In: International Conference on Reversible Computation, (pp. 162–174). Springer International Publishing, Cham (August, 2018)
101. Ifrah, G.: The Universal History of Computing: From the Abacus to Quantum Computing. Wiley (2000)

102. Jain, J., Agrawal, R.: Design and development of efficient reversible floating point arithmetic unit, pp. 811–815 (2015). https://doi.org/10.1109/CSNT.2015.215
103. Jang, K., Choi, S., Kwon, H., Kim, H., Park, J., Seo, H.: Grover on korean block ciphers. Appl. Sci. **10**(18), 6407 (2020). https://doi.org/10.3390/app10186407. https://www.mdpi.com/2076-3417/10/18/6407
104. Jang, K., Choi, S., Kwon, H., Seo, H.: Grover on SPECK: Quantum resource estimates. Cryptology ePrint Archive, Report 2020/640 (2020). https://eprint.iacr.org/2020/640
105. Jang, K., Kim, W., Lim, S., Kang, Y., Yang, Y., Seo, H.: Quantum binary field multiplication with optimized toffoli depth and extension to quantum inversion. Sensors **23**(6), 3156 (2023)
106. Jang, K., Song, G., Kim, H., Kwon, H., Seo, H.: Efficient implementation of present and gift on quantum computers. Appl. Sci. **11**, 4776 (2021). https://doi.org/10.3390/app11114776
107. Jaques, S., Naehrig, M., Roetteler, M., Virdia, F.: Implementing grover oracles for quantum key search on AES and lowmc. In: Canteaut, A., Ishai, Y. (eds.) Advances in Cryptology - EUROCRYPT 2020, Zagreb, Croatia, May 10–14, Part II. Lecture Notes in Computer Science, vol. 12106, pp. 280–310. Springer (2020). https://doi.org/10.1007/978-3-030-45724-2_10
108. Jayashree, H., Thapliyal, H., Arabnia, H.R., Agrawal, V.K.: Ancilla-input and garbage-output optimized design of a reversible quantum integer multiplier. J. Supercomput. **72**, 1477–1493 (2016)
109. Kahanamoku-Meyer, G.D., Yao, N.Y.: Fast quantum integer multiplication with zero ancillas. arXiv preprint arXiv:2403.18006 (2024)
110. Kak, S.C.: Quantum Neural Computing, pp. 259–313. Elsevier (1995). https://doi.org/10.1016/S1076-5670(08)70147-2
111. Karatsuba, A.A., Ofman, Y.P.: Multiplication of many-digital numbers by automatic computers. Doklady Akademii Nauk **145**, 293–294 (1962). Russian Academy of Sciences
112. Katabarwa, A., Gratsea, K., Caesura, A., Johnson, P.D.: Early fault-tolerant quantum computing (2023). https://doi.org/10.48550/ARXIV.2311.14814. https://arxiv.org/abs/2311.14814
113. Kepley, S., Steinwandt, R.: Quantum circuits for $\mathbb{F}_{2^n}$-multiplication with subquadratic gate count. Quantum Inf. Process. **14**, 2373–2386 (2015)
114. Khosropour, A., Aghababa, H., Forouzandeh, B.: Quantum division circuit based on restoring division algorithm. In: 2011 Eighth International Conference on Information Technology: New Generations, pp. 1037–1040. IEEE (2011)
115. Kim, T., Choi, B.S.: Efficient decomposition methods for controlled-rn using a single ancillary qubit. Sci. Rep. **8**(1), 5445 (2018). https://doi.org/10.1038/s41598-018-23764-x
116. Kimmel, S., Lin, C.Y.Y., Lin, H.H.: Oracles with costs. Communication and Cryptography, TQC (2015)
117. Koch, D., Samodurov, M., Projansky, A., Alsing, P.M.: Gate-based circuit designs for quantum adder-inspired quantum random walks on superconducting qubits. Int. J. Quantum Inf. **20**(03), 2150043 (2022). https://doi.org/10.1142/S021974992150043X
118. Kogge, P., Stone, H.: A parallel algorithm for the efficient solution of a general class of recurrence equations. IEEE Trans. Comput. **C-22**, 786–793 (1973). https://doi.org/10.1109/TC.1973.5009159
119. Krinner, S., Lacroix, N., Remm, A., Di Paolo, A., Genois, E., Leroux, C., Hellings, C., Lazar, S., Swiadek, F., Herrmann, J., et al.: Realizing repeated quantum error correction in a distance-three surface code. Nature **605**(7911), 669–674 (2022)
120. Kshirsagar, R.: Robust Quantum Algorithms for Early Fault-Tolerant Quantum Computing. IntechOpen (2024). https://doi.org/10.5772/intechopen.1004416
121. Ladner, R., Fischer, M.: Parallel prefix computation. J. ACM **27**, 831–838 (10 1980). https://doi.org/10.1145/322217.322232
122. Larasati, H.T., Awaludin, A.M., Ji, J., Kim, H.: Quantum circuit design of toom 3-way multiplication. Appl. Sci. **11**(9), 3752 (2021)
123. Lee, I., Lee, K.: The internet of things (iot): Applications, investments, and challenges for enterprises. Bus. Horizons **58**(4), 431–440 (2015)

124. Lee, S., Lee, S.J., Kim, T., Lee, J.S., Biamonte, J., Perkowski, M.: The cost of quantum gate primitives. J. Multiple Valued Logic Soft Comput. **12**, 561–573 (01 2006)
125. Li, S., Song, W., Fang, L., Chen, Y., Ghamisi, P., Benediktsson, J.A.: Deep learning for hyperspectral image classification: An overview. IEEE Trans. Geosci. Remote Sens. **57**(9), 6690–6709 (2019)
126. Lim, S., Kim, H., Jang, K., Wang, S., Baksi, A., Chattopadhyay, A., Seo, H.: Optimized quantum circuit implementation of payoff function. In: 2023 IFIP/IEEE 31st International Conference on Very Large Scale Integration (VLSI-SoC). IEEE (Oct 2023). https://doi.org/10.1109/VLSI-SoC57769.2023.10321843
127. Lin, C.C., Chakrabarti, A., Jha, N.K.: Qlib: Quantum module library. ACM J. Emerg. Technol. Comput. Syst. (JETC) **11**(1), 1–20 (2014)
128. Ling, H.: High speed binary parallel adder. IEEE Trans. Electron. Comput. **EC-15**(5), 799–802 (1966). https://doi.org/10.1109/PGEC.1966.264571
129. Liu, Q., Preneel, B., Zhao, Z., Wang, M.: Improved quantum circuits for AES: Reducing the depth and the number of qubits. Cryptology ePrint Archive, Paper 2023/1417 (2023). https://eprint.iacr.org/2023/1417
130. Liu, X., Yang, H., Yang, L.: Cnot-count optimized quantum circuit of the shor's algorithm. arXiv preprint arXiv:2112.11358 (2021)
131. Liu, Y., Arunachalam, S., Temme, K.: A rigorous and robust quantum speed-up in supervised machine learning. Nat. Phys. **17**(9), 1013–1017 (2021). https://doi.org/10.1038/s41567-021-01287-z
132. Livni, R., Shalev-Shwartz, S., Shamir, O.: On the computational efficiency of training neural networks. In: Advances in Neural Information Processing Systems 27: Annual Conference on Neural Information Processing Systems 2014, December 8–13 2014, Montreal, Quebec, Canada, pp. 855–863 (2014). https://proceedings.neurips.cc/paper/2014/hash/3a0772443a0739141292a5429b952fe6-Abstract.html
133. Lloyd, S., Mohseni, M., Rebentrost, P.: Quantum principal component analysis. Nat. Phys. **10**(9), 631–633 (2014)
134. Low, G.H., Chuang, I.L.: Optimal hamiltonian simulation by quantum signal processing. Phys. Rev. Lett. **118**(1), 010501 (2017). https://doi.org/10.1103/PhysRevLett.118.010501
135. Ma, Y., Roy, S., Miao, J., Chen, J., Yu, B.: Cross-layer optimization for high speed adders: A pareto driven machine learning approach. Trans. Comput. Aided Des. Integr. Circuits Syst. **38**(12), 2298–2311 (2019). https://doi.org/10.1109/TCAD.2018.2878129
136. Markov, I.L., Saeedi, M.: Constant-optimized quantum circuits for modular multiplication and exponentiation. arXiv preprint arXiv:1202.6614 (2012)
137. Maronese, M., Destri, C., Prati, E.: Quantum activation functions for quantum neural networks. Quantum Inf. Process. **21**(4), 128 (2022)
138. Maslov, D.: Reversible logic synthesis benchmarks page (2011). http://webhome.cs.uvic.ca/~dmaslov/
139. Maynard, C.M., Pius, E.: A quantum multiply-accumulator. Quantum Inf. Process. **13**(5), 1127–1138 (2014)
140. McKay, D.C., Hincks, I., Pritchett, E.J., Carroll, M., Govia, L.C., Merkel, S.T.: Benchmarking quantum processor performance at scale. arXiv preprint arXiv:2311.05933 (2023)
141. Merton, R.C.: Theory of rational option pricing. Bell J. Econom. Manag. Sci. **4**(1), 141 (1973). https://doi.org/10.2307/3003143
142. Moore, G.E.: Cramming more components onto integrated circuits. Proc. IEEE **86**(1), 82–85 (1998)
143. Muñoz-Coreas, E., Thapliyal, H.: Quantum circuit design of a t-count optimized integer multiplier. IEEE Trans. Comput. **68**(5), 729–739 (2018)
144. Nam, Y., Su, Y., Maslov, D.: Approximate quantum fourier transform with o(n log(n)) t gates. NPJ Quantum Inf. **6**(1), (2020). https://doi.org/10.1038/s41534-020-0257-5
145. Nguyen, A.Q.: Optimal reversible quantum circuit for multiplication. Technical Report TR-2004007, City University of New York (CUNY) (2004)

146. Nguyen, Q.T., Schatzki, L., Braccia, P., Ragone, M., Coles, P.J., Sauvage, F., Larocca, M., Cerezo, M.: Theory for equivariant quantum neural networks. PRX Quantum **5**(2), 020328 (2024). https://doi.org/10.1103/PRXQuantum.5.020328
147. Nguyen, T., Van Meter, R.: A space-efficient design for reversible floating point adder in quantum computing. ACM J. Emerg. Technol. Comput. Syst. **11**, 1–18 (2013). https://doi.org/10.1145/2629525
148. Nie, D., Trullo, R., Lian, J., Wang, L., Petitjean, C., Ruan, S., Wang, Q., Shen, D.: Medical image synthesis with deep convolutional adversarial networks. IEEE Trans. Biomed. Eng. **65**(12), 2720–2730 (2018)
149. Nie, J., Zhu, Q., Li, M., Sun, X.: Quantum circuit design for integer multiplication based on schönhage-strassen algorithm. IEEE Trans. Comput. Aided Des. Integr. Circuits Syst. 4791–4802 (2023)
150. Nielsen, M.A., Chuang, I.L.: Quantum Computation and Quantum Information: 10th Anniversary Edition. Cambridge University Press (2012). https://doi.org/10.1017/CBO9780511976667
151. NIST.: Submission requirements and evaluation criteria for the post-quantum cryptography standardization process (2016). https://csrc.nist.gov/CSRC/media/Projects/Post-Quantum-Cryptography/documents/call-for-proposals-final-dec-2016.pdf
152. NIST.: Call for additional digital signature schemes for the post-quantum cryptography standardization process (2022). https://csrc.nist.gov/csrc/media/Projects/pqc-dig-sig/documents/call-for-proposals-dig-sig-sept-2022.pdf
153. OpenAI(2023): Gpt-4 technical report. arXiv preprint arXiv:2303.08774 (2023)
154. Orts, F., Filatovas, E., Ortega, G., SanJuan-Estrada, J., Garzón, E.: Improving the number of t gates and their spread in integer multipliers on quantum computing. Phys. Rev. A **107**(4), 042621 (2023)
155. Orts, F., Paulavičius, R., Filatovas, E.: Quantum circuit optimization of an integer divider. J. Syst. Software, 112091 (2024)
156. Orts Gómez, F. J., Ortega López, G., Combarro, E., Martín Garzón, G. E.: A review on reversible quantum adders. J. Network Comput. Appl. **170**, 102810 (2020). https://doi.org/10.1016/j.jnca.2020.102810
157. Orús, R., Mugel, S., Lizaso, E.: Quantum computing for finance: Overview and prospects. Rev. Phys. **4**, 100028 (2019). https://doi.org/10.1016/j.revip.2019.100028
158. Paler, A.: Quantum fourier addition simplified to toffoli addition. Phys. Rev. A **106**(4), (2022). https://doi.org/10.1103/PhysRevA.106.042444
159. Panahi, M.M., Hashemipour, O., Navi, K.: A novel design of a multiplier using reversible ternary gates. IETE J. Res. **67**(6), 744–753 (2021)
160. Parent, A., Roetteler, M., Mosca, M.: Improved reversible and quantum circuits for karatsuba-based integer multiplication. arXiv preprint arXiv:1706.03419 (2017)
161. Park, B., Ahn, D.: Reducing cnot count in quantum fourier transform for the linear nearest-neighbor architecture. Sci. Rep. **13**(1), 8638 (2023). https://doi.org/10.1038/s41598-023-35625-3
162. Park, B., Doyeol, A.: T count optimization of approximate quantum fourier transform (2022). https://doi.org/10.48550/ARXIV.2203.07739. https://arxiv.org/abs/2203.07739
163. de Paula Neto, F.M., Ludermir, T.B., de Oliveira, W.R., da Silva, A.J.: Implementing any nonlinear quantum neuron. IEEE Trans. Neural Networks Learn. Syst. **31**(9), 3741–3746 (2019)
164. Pavlidis, A., Floratos, E.: Quantum-fourier-transform-based quantum arithmetic with qudits. Phys. Rev. A **103**(3), 032417 (2021). https://doi.org/10.1103/PhysRevA.103.032417
165. Pavlidis, A., Gizopoulos, D.: Fast quantum modular exponentiation architecture for shor's factorization algorithm (2012). https://doi.org/10.48550/ARXIV.1207.0511. https://arxiv.org/abs/1207.0511
166. Peral-García, D., Cruz-Benito, J., García-Peñalvo, F.J.: Systematic literature review: Quantum machine learning and its applications. Comput. Sci. Rev. **51**, 100619 (2024)
167. Preskill, J.: Quantum computing in the nisq era and beyond. Quantum **2**, 79 (2018)

168. Putranto, D.S.C., Wardhani, R.W., Larasati, H.T., Kim, H.: Space and time-efficient quantum multiplier in post quantum cryptography era. IEEE Access **11**, 21848–21862 (2023)
169. Rajendran, G., Basak, D., Deb, S., Wang, S., Chattopadhyay, A.: A novel current comparator enabling large rram crossbars for bnns and pufs. In: 2024 IFIP/IEEE 32nd International Conference on Very Large Scale Integration (VLSI-SoC), pp. 1–6. IEEE (2024)
170. Ramezani, M., Nikaeen, M., Farman, F., Ashrafi, S.M., Bahrampour, A.: Quantum multiplication algorithm based on the convolution theorem. Phys. Rev. A **108**(5), 052405 (2023)
171. Rebentrost, P., Gupt, B., Bromley, T.R.: Quantum computational finance: Monte carlo pricing of financial derivatives. Phys. Rev. A **98**(2), 022321 (2018). https://doi.org/10.1103/PhysRevA.98.022321
172. Rebentrost, P., Lloyd, S.: Quantum computational finance: quantum algorithm for portfolio optimization (2018). https://doi.org/10.48550/ARXIV.1811.03975. https://arxiv.org/abs/1811.03975
173. Rebentrost, P., Luongo, A., Bosch, S., Lloyd, S.: Quantum computational finance: martingale asset pricing for incomplete markets (2022). https://doi.org/10.48550/ARXIV.2209.08867. https://arxiv.org/abs/2209.08867
174. Rebentrost, P., Mohseni, M., Lloyd, S.: Quantum support vector machine for big data classification. Phys. Rev. Lett. **113**(13), 130503 (2014)
175. Rines, R., Chuang, I.: High performance quantum modular multipliers. arXiv preprint arXiv:1801.01081 (2018)
176. Rosenberger, G.B.: Simultaneous carry adder. US Patent 2,966, vol. 30 (1957)
177. Ruiz-Perez, L., Garcia-Escartin, J.C.: Quantum arithmetic with the quantum fourier transform. Quantum Inf. Process. **16**(6), (2017). https://doi.org/10.1007/s11128-017-1603-1
178. Gayathri, S.S., Kumar, R., Haghparast, M., Samiappan, D.: A novel and efficient square root computation quantum circuit for floating-point standard. Int. J. Theor. Phys. **61**, 1–15 (2022). https://doi.org/10.1007/s10773-022-05222-7
179. Gayathri, S.S., Kumar, R., Samiappan, D., Kaushik, B.K.: T-count optimized quantum circuit for floating point addition and multiplication. Quantum Inf. Process. **20**, 378 (2021). https://doi.org/10.1007/s11128-021-03296-6
180. Gayathri, S.S., Kumar, R., Samiappan, D., Kaushik, B.K., Haghparast, M.: T-count optimized wallace tree integer multiplier for quantum computing. Int. J. Theor. Phys. **60**, 1–13 (2021). https://doi.org/10.1007/s10773-021-04864-3
181. Sajadimanesh, S., Atoofian, E.: Implementation of a quantum division circuit on noisy intermediate-scale quantum devices using dynamic circuits and approximate computing. Phys. Rev. A **109**, 052601 (2024). https://doi.org/10.1103/PhysRevA.109.052601. https://link.aps.org/doi/10.1103/PhysRevA.109.052601
182. Sajadimanesh, S., Faye, J.P.L., Atoofian, E.: Nisq-friendly non-linear activation functions for quantum neural networks. In: 2022 IEEE International Conference on Networking, Architecture and Storage (NAS). IEEE (Oct 2022). https://doi.org/10.1109/NAS55553.2022.9925456
183. Sajadimanesh, S., Faye, J.P.L., Atoofian, E.: Practical approximate quantum multipliers for nisq devices. In: Proceedings of the 19th ACM International Conference on Computing Frontiers, pp. 121–130 (2022)
184. Schönhage, A., Strassen, V.: Fast multiplication of large numbers. Computing **7**, 281–292 (1971)
185. Schuld, M., Sinayskiy, I., Petruccione, F.: The quest for a quantum neural network. Quantum Inf. Process. **13**, 2567–2586 (2014)
186. Schuld, M., Sinayskiy, I., Petruccione, F.: An introduction to quantum machine learning. Contemp. Phys. **56**(2), 172–185 (2015)
187. Schuld, M., Sinayskiy, I., Petruccione, F.: Simulating a perceptron on a quantum computer. Phys. Lett. A **379**(7), 660–663 (2015)
188. Seidel, R., Tcholtchev, N., Bock, S., Becker, C.K.U., Hauswirth, M.: Efficient floating point arithmetic for quantum computers. IEEE Access **10**, 72400–72415 (2022). https://doi.org/10.1109/ACCESS.2022.3188251

189. Selinger, P.: Quantum circuits of t-depth one. Phys. Rev. A **87**, 042302 (2013). https://doi.org/10.1103/PhysRevA.87.042302. https://link.aps.org/doi/10.1103/PhysRevA.87.042302
190. Sharma, S., Sharma, S., Athaiya, A.: Activation functions in neural networks. Towards Data Sci. **6**(12), 310–316 (2017)
191. Shaw, R.F.: Arithmetic operations in a binary computer. Rev. Sci. Instrum. **21**(8), 687–693 (1950)
192. Shi, H., Feng, X.: Quantum circuits of AES with a low-depth linear layer and a new structure. Cryptology ePrint Archive, Paper 2024/381 (2024). https://eprint.iacr.org/2024/381
193. Shor, P.: Algorithms for quantum computation: Discrete logarithms and factoring. In: Proceedings of 35th Annual Symposium on Foundations of Computer Science (10 1996). https://doi.org/10.1109/SFCS.1994.365700
194. Singleton Jr, R.L.: Shor's factoring algorithm and modular exponentiation operators: A pedagogical presentation with examples. arXiv preprint arXiv:2306.09122 (2023)
195. Sklansky, J.: Conditional-sum addition logic. IRE Trans. Electron. Comput. **EC-9**, 226–231 (1960). https://doi.org/10.1109/TEC.1960.5219822
196. Smolin, J.A., Smith, G., Vargo, A.: Oversimplifying quantum factoring. Nature **499**(7457), 163–165 (2013)
197. Song, M.h., Jang, K.b., Song, G.j., Kim, W.w., Seo, H.J.: Quantum circuit implementation of the led block cipher with compact qubit. J. Korea Inst. Inf. Secur. Cryptol. **33**(3), 383–389 (2023)
198. Sousa, L.: Nonconventional computer arithmetic circuits, systems and applications. IEEE Circuits Syst. Mag. **21**(1), 6–40 (2021). https://doi.org/10.1109/MCAS.2020.3027425
199. Stamatopoulos, N., Egger, D.J., Sun, Y., Zoufal, C., Iten, R., Shen, N., Woerner, S.: Option pricing using quantum computers. Quantum **4**, 291 (2020). https://doi.org/10.22331/q-2020-07-06-291
200. Stamatopoulos, N., Zeng, W.J.: Derivative pricing using quantum signal processing. Quantum **8**, 1322 (2024). https://doi.org/10.22331/q-2024-04-30-1322
201. Steiger, D., Häner, T., Troyer, M.: Projectq: An open source software framework for quantum computing. Quantum **2**, 49 (2016). https://doi.org/10.22331/q-2018-01-31-49
202. Sünderhauf, C., Campbell, E., Camps, J.: Block-encoding structured matrices for data input in quantum computing. Quantum **8**, 1226 (2024). https://doi.org/10.22331/q-2024-01-11-1226
203. Tacchino, F., Macchiavello, C., Gerace, D., Bajoni, D.: An artificial neuron implemented on an actual quantum processor. NPJ Quantum Inf. **5**(1), 26 (2019)
204. Takahashi, Y.: Quantum arithmetic circuits: a survey. IEICE Trans. Fund. Electron. Commun. Comput. Sci. **92**(5), 1276–1283 (2009)
205. Takahashi, Y., Kunihiro, N.: A linear-size quantum circuit for addition with no ancillary qubits. Quantum Info. Comput. **5**(6), 440–448 (2005)
206. Takahashi, Y., Kunihiro, N.: A quantum circuit for shor's factoring algorithm using 2n + 2 qubits. Quantum Info. Comput. **6**(2), 184–192 (2006)
207. Takahashi, Y., Kunihiro, N.: A fast quantum circuit for addition with few qubits. Quantum Inf. Comput. **8**, 636–649 (2008). https://doi.org/10.26421/QIC8.6-7-5
208. Takahashi, Y., Tani, S., Kunihiro, N.: Quantum addition circuits and unbounded fan-out. Quantum Inf. Comput. **10**(9), 872–890 (2010)
209. Thapliyal, H.: Mapping of subtractor and adder-subtractor circuits on reversible quantum gates. Trans. Comput. Sci. **XXVII**, 10–34 (2016). Springer
210. Thapliyal, H., Munoz-Coreas, E., Varun, T., Humble, T.S.: Quantum circuit designs of integer division optimizing t-count and t-depth. IEEE Trans. Emerg. Top. Comput. **9**(2), 1045–1056 (2019)
211. Thapliyal, H., Ranganathan, N.: Design of efficient reversible binary subtractors based on a new reversible gate. In: 2009 IEEE Computer Society Annual Symposium on VLSI, pp. 229–234. IEEE (2009)
212. Tilly, J., Chen, H., Cao, S., Picozzi, D., Setia, K., Li, Y., Grant, E., Wossnig, L., Rungger, I., Booth, G.H., et al.: The variational quantum eigensolver: a review of methods and best practices. Phys. Rep. **986**, 1–128 (2022)

213. Udvarnoki, Z., Fáth, G., Fogarasi, N.: Quantum advantage of monte carlo option pricing. J. Phys. Commun. **7**(5), 055001 (2023). https://doi.org/10.1088/2399-6528/acd2a4
214. van Dam, W., Hallgren, S.: Efficient quantum algorithms for shifted quadratic character problems (2001). https://arxiv.org/abs/quant-ph/0011067
215. van Dam, W., Hallgren, S., Ip, L.: Quantum algorithms for some hidden shift problems. SIAM J. Comput. **36**, 763–778 (2002). https://doi.org/10.1145/644108.644189
216. van Hoof, I.: Space-efficient quantum multiplication of polynomials for binary finite fields with sub-quadratic toffoli gate count. arXiv preprint arXiv:1910.02849 (2019)
217. Van Meter, R., Itoh, K.M.: Fast quantum modular exponentiation. Phys. Rev. A **71**(5), 052320 (2005). https://doi.org/10.1103/PhysRevA.71.052320
218. Vedral, V., Barenco, A., Ekert, A.: Quantum networks for elementary arithmetic operations. Phys. Rev. A **54**, 147–153 (1995). https://doi.org/10.1103/PhysRevA.54.147
219. Von Neumann, J., Burks, A.W., et al.: Theory of self-reproducing automata. IEEE Trans. Neural Networks **5**(1), 3–14 (1966)
220. Wack, A., Paik, H., Javadi-Abhari, A., Jurcevic, P., Faro, I., Gambetta, J.M., Johnson, B.R.: Quality, speed, and scale: three key attributes to measure the performance of near-term quantum computers. arXiv preprint arXiv:2110.14108 (2021)
221. Wallace, C.S.: A suggestion for a fast multiplier. IEEE Trans. Electron. Comput. (1), 14–17 (1964)
222. Wan, K.H., Dahlsten, O., Kristjánsson, H., Gardner, R., Kim, M.: Quantum generalisation of feedforward neural networks. NPJ Quantum Inf. **3**(1), 36 (2017)
223. Wang, F., Luo, M., Li, H., Qu, Z., Wang, X.: Improved quantum ripple-carry addition circuit. Sci. China Inf. Sci. **59**, 042406 (2016). https://doi.org/10.1007/s11432-015-5411-x
224. Wang, S., Baksi, A., Chattopadhyay, A.: A higher radix architecture for quantum carry-lookahead adder. Sci. Rep. **13**, 16338 (2023). https://doi.org/10.1038/s41598-023-41122-4
225. Wang, S., Chattopadhyay, A.: Efficient depth optimization in quantum addition and modular arithmetic with ling structure. In: IFIP/IEEE International Conference on Very Large Scale Integration-System on a Chip, pp. 73–89. Springer (2023)
226. Wang, S., Chattopadhyay, A.: Reducing depth of quantum adder using ling structure. In: VLSI-SoC (2023)
227. Wang, S., Jang, K., Baksi, A., Chakraborty, S., Lee, B., Chattopadhyay, A., Seo, H.: New results in quantum analysis of led: Featuring one and two oracle attacks. Cryptology ePrint Archive (2024)
228. Wang, S., Jin, A., Deb, S., Dutta, T., Mukherjee, M., Chattopadhyay, A.: Poster: Malaq-a malware against quantum computer. In: Proceedings of the 19th ACM Asia Conference on Computer and Communications Security, pp. 1946–1948 (2024)
229. Wang, S., Li, X., Lee, W.J.B., Deb, S., Lim, E., Chattopadhyay, A.: A comprehensive study of quantum arithmetic circuits. Phil. Trans. A **383**(2288), 20230392 (2025)
230. Wang, S., Lim, E., Chattopadhyay, A.: Boosting the efficiency of quantum divider through effective design space exploration. arXiv preprint arXiv:2403.01206 (2024)
231. Wang, S., Lim, E., Li, X., Feng, J., Chattopadhyay, A.: Minimum depth quantum modular addition through carry-save architecture. In: 2024 IFIP/IEEE 32nd International Conference on Very Large Scale Integration (VLSI-SoC), pp. 1–6. IEEE (2024)
232. Wang, S., Mondal, A., Chattopadhyay, A.: Optimal toffoli-depth quantum adder. ACM Trans. Quantum Comput. (2025). https://doi.org/10.1145/3743691. https://doi-org.remotexs.ntu.edu.sg/10.1145/3743691
233. Wiebe, N., Roetteler, M.: Quantum arithmetic and numerical analysis using repeat-until-success circuits. arXiv preprint arXiv:1406.2040 (2014)
234. Wilde, M.M.: Quantum-shift-register circuits. Phys. Rev. A **79**(6), 062325 (2009). https://doi.org/10.1103/PhysRevA.79.062325
235. Wu, L., Zhang, X., Wang, S., Hu, W.: Hardware trojan detection at lut: Where structural features meet behavioral characteristics. In: 2022 IEEE International Symposium on Hardware Oriented Security and Trust (HOST), pp. 121–124. IEEE (2022)

236. Xiang, Z., Zeng, X., Lin, D., Bao, Z., Zhang, S.: Optimizing implementations of linear layers. IACR Trans. Symmetric Cryptol. 120–145 (2020)
237. Yalovetzky, R., Minssen, P., Herman, D., Pistoia, M.: Hybrid hhl with dynamic quantum circuits on real hardware (2021). https://doi.org/10.48550/ARXIV.2110.15958. https://arxiv.org/abs/2110.15958
238. Yan, S., Qi, H., Cui, W.: Nonlinear quantum neuron: A fundamental building block for quantum neural networks. Phys. Rev. A **102**(5), 052421 (2020)
239. Yao, J., Erichson, N.B., Lopes, M.E.: Error estimation for random fourier features (2023). https://doi.org/10.48550/ARXIV.2302.11174. https://arxiv.org/abs/2302.11174
240. Yi, X., Babyn, P.: Sharpness-aware low-dose ct denoising using conditional generative adversarial network. J. Digit. Imag. **31**, 655–669 (2018)
241. Young, R.C., Birch, P.M., Chatwin, C.R.: A simplification of the shor quantum factorization algorithm employing a quantum hadamard transform. In: Pattern Recognition and Tracking XXIX, vol. 10649, pp. 12–21. SPIE (2018)
242. Yuan, S., Gao, S., Wen, C., Wang, Y., Qu, H., Wang, Y.: A novel fault-tolerant quantum divider and its simulation. Quantum Inf. Process. **21**(5), 182 (2022)
243. Yuan, Y., Wang, C., Wang, B., Chen, Z.Y., Dou, M.H., Wu, Y.C., Guo, G.P.: An improved qft-based quantum comparator and extended modular arithmetic using one ancilla qubit. New J. Phys. **25**(10), 103011 (2023). https://doi.org/10.1088/1367-2630/acfd52
244. Yuan, Y., Wu, W., Shi, T., Zhang, L., Zhang, Y.: A framework to improve the implementations of linear layers. IACR Trans. Symmet. Cryptol. **2024**(2), 322–347 (2024). https://tosc.iacr.org/index.php/ToSC/article/view/11633
245. Zalka, C.: Fast versions of shor's quantum factoring algorithm. arXiv preprint quant-ph/9806084 (1998)
246. Zhan, J.: Quantum multiplier based on exponent adder. arXiv preprint arXiv:2309.10204 (2023)
247. Zhang, A., Wang, X., Zhao, S.: The multiplier based on quantum fourier transform. CCF Trans. High Perform. Comput. **2**, 221–227 (2020)
248. Zhang, Y.: Four arithmetic operations on the quantum computer. J. Phys. Confer. Ser. **1575**, 012037 (2020). https://doi.org/10.1088/1742-6596/1575/1/012037
249. Zhao, J., Zhang, Y.H., Shao, C.P., Wu, Y.C., Guo, G.C., Guo, G.P.: Building quantum neural networks based on a swap test. Phys. Rev. A **100**(1), 012334 (2019)
250. Zhao, S., Li, H., Li, G., Tang, X.: The implementation of the enhanced quantum floating-point adder. Modern Phys. Lett. A **37**, 2250169 (2022). https://doi.org/10.1142/S0217732322501693
251. Zhou, L., Wang, S.T., Choi, S., Pichler, H., Lukin, M.D.: Quantum approximate optimization algorithm: Performance, mechanism, and implementation on near-term devices. Phys. Rev. X **10**(2), 021067 (2020)
252. Zi, W., Wang, S., Kim, H., Sun, X., Chattopadhyay, A., Rebentrost, P.: Efficient quantum circuits for machine learning activation functions including constant t-depth relu (2024). https://doi.org/10.48550/ARXIV.2404.06059. https://arxiv.org/abs/2404.06059
253. Çkmak, S., Kurt, M., Gençten, A.: Qft based quantum arithmetic logic unit on ibm quantum computer (2023). https://doi.org/10.48550/ARXIV.2306.09560. https://arxiv.org/abs/2306.09560
254. Şahin, E.: Quantum arithmetic operations based on quantum fourier transform on signed integers. Int. J. Quantum Inf. **18**(06), 2050035 (2020). https://doi.org/10.1142/S0219749920500355

MIX
Papier aus verantwortungsvollen Quellen
Paper from responsible sources
FSC® C105338

If you have any concerns about our products,
you can contact us on
ProductSafety@springernature.com

In case Publisher is established outside the EU,
the EU authorized representative is:
Springer Nature Customer Service Center GmbH
Europaplatz 3, 69115 Heidelberg, Germany

Printed by Libri Plureos GmbH
in Hamburg, Germany